# UNDERWATER PARADISE

**RED SEA, THE MALDIVES, MALAYSIA AND THE CARIBBEAN**

WHITE STAR PUBLISHERS

# UNDERWATER PARADISE

**RED SEA, THE MALDIVES, MALAYSIA AND THE CARIBBEAN**

Texts
Angelo Mojetta
Andrea and
Antonella Ferrari

Illustrations
Monica Falcone

Translations
Antony Shugaar
A.B.A. s.r.l., Milan
Studio Traduzioni
Vecchia, Milan
Barbara Fisher

# Contents

INTRODUCTION     PAGE    4

THE RED SEA     PAGE    14

THE MALDIVES     PAGE   108

MALAYSIA     PAGE   202

THE CARIBBEAN     PAGE   294

*1 A large grouper (Epinephelus tauvina) peeks out from among hard coral formations typical of the Caribbean.*

*2-3 A group of anthias swim around a hard coral formation covered with anemones, sea lilies and sponges.*

*4 left A squirrel fish (Holocentrus ascensionis) menacingly faces a photographer.*

*4 right This hard coral formation has been colonised by sponges and sea fans.*

*5 A dense shoal of jacks (Caranx sp.) swims in their typical circular formation creating a silver glow all around.*

*© 2000, 2008 White Star S.p.A.*
*Via Candido Sassone, 22/24*
*13100 Vercelli, Italy*
*www.whitestar.it*

*ISBN: 978-88-544-0333-8*

2 3 4 5 6  12 11 10 09 08

Printed in Indonesia.

# INTRODUCTION

Aptly called "underwater paradises", these fantastic submarine worlds that are quite real and owe nothing to fantasy, ignites the imagination, not only of divers and scuba enthusiasts, but also of just about anyone who peruses these pages. Genesis tells us that the waters were created the third day, but Man came a little later. Man was born of the earth and is therefore not an aquatic creature, although 70% of the human body is water. Man, in fact, starts off his life fishlike, in the aquatic environment of his mother's womb, to develop into a land-bound creature. It is perhaps because of our aquatic origin that a passion for diving seems to be spreading like a sort of fever, infecting thousands of new enthusiasts each year.

While the Garden of Eden, that we have incessantly searched for on Earth after having lost it, is by definition a land-based environment, all its features can be found under the surface of the seas in a world that we have only recently discovered, but towards which we find ourselves almost irresistibly drawn since it abounds in all that was lost when Man was driven out of Paradise: calm, warmth, extraordinary colors, life, in all its splendour, surrounding us, indeed embracing us, in a manner so difficult to find on land. Besides all these extraordinary attractions, one must not forget that below the surface we are weightless and can hover around almost effortlessly in a fantastic three-dimensional world, enjoying the thrilling sensations that perhaps only birds can feel. Anyone diving at the sites dealt with here, from the Red Sea to Malaysia to the Caribbean will be sure to fall under a sort of spell that will endure long after the dive since it is the result of the peculiar sensations that all divers feel but find difficult to describe. Diving a reef is tantamount to entering a world where everything is color, shape and movement. Even the

most experienced marine biologists, more familiar than others with these enchanting sites, at a certain time break off their professional carapace to abandon themselves unrepenting to the extraordinary feeling of being steeped in the midst of innumerable life forms that here, more than in any other place, become commonplace reality.

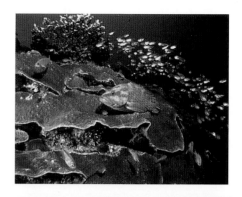

These seemingly infinite underwater life forms also come in an infinite variety of forms: fine filaments of whip-like coral, black coral urchins, sponges that seem like organ pipes or barrels, round, oval, tubular, cubic and flat fish, sea fans, dense forests of soft or hard coral moulded into their thousand shapes by the knowing hand of the sea. The main problem here is to focus on a particular subject. While with effort, this may be possible for a while, it is

the subject itself that will lead the observer elsewhere. A platform of acropora hides a blue and yellow ray that rises almost as in flight, from the seabed to cross the trajectory of a shoal of batfish that suddenly disperses with the arrival of a shoal of jacks or barracuda that improvise one of their mysterious twirling arabseques before disappearing into the deep blue to allow the diver to distinguish the streamlined silhouette of a shark that that powerfully skims just above the acropora from which the whole scene started off. There is no beginning and no end to this voyage of discovery and one's very first dive into these underwater paradises is an almost infinite experience. In fact, not only is each sea different from all the others, but each dive in itself is a unique experience since there will always be that first wave, that first fish (there is a first time for each species that the diver

meets and learns to recognize), the first cowry, making each dive, a "first" dive. The common backdrop to all this is the salt water world, a chemical combination of water and chemical elements that many have tried to define without ever fully succeeding. Considered by some to be an intermediate world between the divine and the human, the sea is a sort of fourth dimension in which sensations frequently prevail even if the

time spent underwater is limited. Down time is strictly regulated by unbending rules, in function of remaining air reserves, arid mathematical tables and the changing figures on the screen of the underwater computer. Immense space is at once foreign to us and yet part of us, and one is bound to feel the same overwhelming sensations whether one dives into the seemingly unbounded space of the submarine world or contemplates our tiny blue planet from a spaceship. This is a world that does not tick to our pace but that is in tune with the entire universe, whose timeless rhythm and pulse is

*6 top
A blacktip grouper (Epinephelus fasciatus) lies in wait of prey on a coral formation.*

*6-7 Innumerable scalefin anthias (Pseudanthias squamipinnis), with their typical orange hue, swim close to large. Lively-colored alcyonnarians.*

*7 top A masked puffer (Arothron diadematus) tries to camouflage itself against a coral block.*

*7 center This picture brings out the extraordinary colors of a Queen angelfish (Holacantus ciliaris), typically found in the Caribbean.*

*7 bottom This sea fan is temporarily colonized by a few sea lilies.*

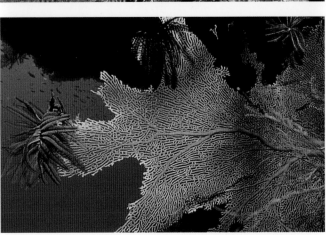

8 A coral hind
(Cephalopholis
miniata) *with its
jaws wide open
waits in ambush
for unsuspecting
prey.*

9 top *A terrible reef
stonefish*
(Synanceia
verrucosa) *parks
itself on a coral
formation. This fish
is a master of
camouflage and
can only be
recognized by its
arch-shaped
mouth.*

perhaps most noticeable to us in the fascinating regularity of the ebb and flow of the tides.

The sea, therefore, at once attracts and repels us humans, and to make it more familiar or at least to create a semblance of being more at ease and not too alone in so much beauty, we have been unable to do more than to name its lifeforms after familiar creatures. This is why below the surface, we find sea hounds and catfish, butterflyfish and angelfish, flower animals, sea serpents, seahens, surgeonfish, sea anemones, toadfish, leaffish, filefish, hammerhead sharks, starfish, sea urchins, sea hares, sea snails, damselfish, sea thrushes and eaglefish, all named after birds, animals, humans or even fantastic creatures that are part of our land bound world, in an attempt perhaps to ensure that our world remains the measure of all things, even under the sea.

But the sea is too vast and immense to remain imprisoned within the bounds of reason. The sea continues to remain a source of mystery and legend, peopled by fantastic creatures. These legends of the sea are still very much alive and provide inspiration to poets when they find the muses above land lacking. The Persian poet Nezami of Ganjé for instance, described the atmosphere of star-studded night by comparing it with the black mouth of a whale full of pearls. There is no doubt however, that despite sea-side colors, scents and the eternal, indefatigable, breaking of the waves, Man remains most fascinated by the seas and oceans because of the extraordinary life forms, some of which are unique, that populate their waters.

The close interdependence that develops between various organisms and their surrounding environment gives rise to ecologists call an ecosystem, that is to say, a part of the world divisible into individual elements that, almost as though they make up a colossal jig-saw puzzle, mesh together and are all necessary for each of the elements as well as the whole system, to function.

It is sometimes difficult for a layman to understand the importance and the effect of such factors, and yet, when there is widespread understanding of the waft that brings these subtle inter-relationships together, Man's view of the sea is bound to change drastically because we will suddenly be faced with the real effects of all of our actions.

The next step, as often happens after insight, is only to fine tune our knowledge. But that is another story altogether.

## MAN IS A PART OF EVERYTHING

To understand the complexity of forms that separate the simplest living organism from the biosphere of which it is a part, one must first understand a whole hierarchical scale of which Man is also an element by the very fact of his existence. Nothing on the scale is superfluous and no action remains without a reaction. The very fact that we humans have been able to enter the underwater world is sufficient to generate a small environmental revolution that affects millions of beings, without our even realising it. The same things happens,

perhaps, when a humpback whale breaks through the surface of the sea to splash back with the full weight of its body, a game or show of force that these cetaceans also use as an efficient system of communication that Man can only try to imitate. The levels of this hierarchy are rigidly fixed, much more so in fact, than in any ordered structure that Man has been able to devise and even if each hierarchical level can be studied separately, one will always, inevitably reach a point at which one must necessarily study another successive or previous link in the chain in order to fully understand what one has just discovered.
A diver calmly using his flippers

to swim leisurely above a reef, a favourite pass-time for tens of thousands of diving enthusiasts, would most probably be blissfully unaware of the extent to which the variety and beauty of these seabed are the fruit of a complex series of relationships that involves local fish, invertebrates and corals whose well-being depends on minute algae, known as zooxanthellae, that in turn depend on light. A reef, however, is infinitely more than this. These small patches of paradise can, in fact, develop only if the surrounding waters clear and warm. This vital requirement depends on a vast number of factors, involving minute organisms measuring just a few microns to phenomena on a

global scale, such as the thousands of square meters of the earth's surface covered by water and affected by large-scale oceanic variations in waves, currents, tides, water density, light penetration, temperature, productivity. Water transparency in turn is linked to the concentration of plankton that depends on the availability of nutrients that largely depend on substances poured into the oceans by rivers and rainfall. While temperatures are most favourable for reefs in the tropics, their demarcations are far from being as clearly defined as in our maps, because of hot and cold ocean currents that create substantial differences between areas, sometimes making it possible for a reef to develop in improbable geographical areas and at others, hindering reef development in areas that, according to our maps, would provide ideal conditions for reef development, as in the Galapagos, for instance, where coral reefs are replaced by penguins and sea lions.

*9 bottom left A close-up picture showing the unique shape of the snout of the trumpetfish (Aulostomus maculatus).*

*9 bottom right This giant moray (Gymnothorax javanicus) enjoys the "beauty treatment" as a few tiny prawns free it from irritating parasites.*

## THE FUTURE OF UNDERWATER PARADISES

The newly developed technology that allows Man to visually dominate the oceans, at least in the upper reaches of the water column, may soon play a key role in helping Man understand the inherent fragility of the marine ecosystem that may paradoxically depend on the very vastness of the oceans. The oceans, in fact, seem almost immune to the changes that affect our planet in the short term. Earthquakes, hurricanes, floods, draughts, volcanic eruptions, that have devastating effects on the group, do not seem to leave obvious traces under the sea, almost as if they were so many sand-castles that a few waves soon erase to bring the water edge back to its previous state. The sea's capability to undertake this depends on the incessant re-mixing of sea water, a slow process that can take hundreds, if not thousands of years, but through which the waters at the deep end of the water column are brought to the surface and vice versa. While our biological life-span is too brief for us to realize the effects of phenomena that are not limited to rather narrowly defined areas, current knowledge is deep enough to provide us with insight into how the oceans are vulnerable to large-scale global variations brought about by factors such as pollution and global climate change.

Each year, thousands of new chemical compounds are launched on our markets and eventually end up in the sea. While the fate of some of these may be known and the short-term effects of others may be understood, we are aware of the environmental impact of but a tiny fraction of these new chemicals compounds. We are especially in the dark about how each product may change its behaviour when placed in

10 top
A dense shoal of silver colored jacks seem to block out the sunrays that filter through to the deep.

10-11
A threatening blackfin shark (Carcharinus wheeleri) scouts the sandy seabed in search of prey.

contact with others. Water has always been considered the best possible solvent. This means that the sea may become a huge laboratory flask with a high risk of unforeseeable reactions leading to unimaginable results. The well-known episode of Minamata Bay in Japan that revealed the toxic effects of methyl mercury on Man, is repeated on a daily basis in the case of innumerable marine organisms, from the tiniest, such as plankton and the larvae of a large number of species to marine birds, sharks and the huge cetaceans, that quite like Man, are the top of the marine food chain and therefore condemned to sooner or later accumulate concentrations of toxins in their tissues.

Even temperature changes will have tangible effects of the future life of the oceans. The fact that the temperature of the oceans fluctuates to far lesser degree than the temperature above the surface has resulted

upheavals that have a tangible impact on the ecological features of vast areas of ocean as well as the waters around islands and along coastlines, with devastating effects on a large number of marine species. By pondering over these reflections while perusing the photographs that illustrate these underwater paradises, it is easy to understand why the concept of the marine ecosystem is slowly taking root and developing in our collective conscience. Light, temperature, currents, nutrients and depth all

in the evolution of a large number of marine organisms that are incapable of regulating their body temperature. As a result of this, the physiological processes of these organisms are very delicate, and can function only within very limited temperature ranges, as in the case of the corals, referred to above. Periodical, short-term changes that however affect large areas, such as the El Niño phenomenon can generate full-fledged environmental

come together to form a wondrous mosaic and Man must learn how to keep each tessera in place to preserve the majestic beauty of the whole. It is up to each individual diver to choose whether to be a mere ignorant spectator of nature's

masterpiece or to play an active and responsible role in preserving it. Whatever role individual divers may choose to play with regard to the sea, they must never forget that Man is part and parcel of the oceans and that Man's deliberate choices or indifference are bound to have tremendous impact on marine life, even if these choices are in fact exercised in boardrooms and living rooms seemingly far away from the underwater paradises of the sea.

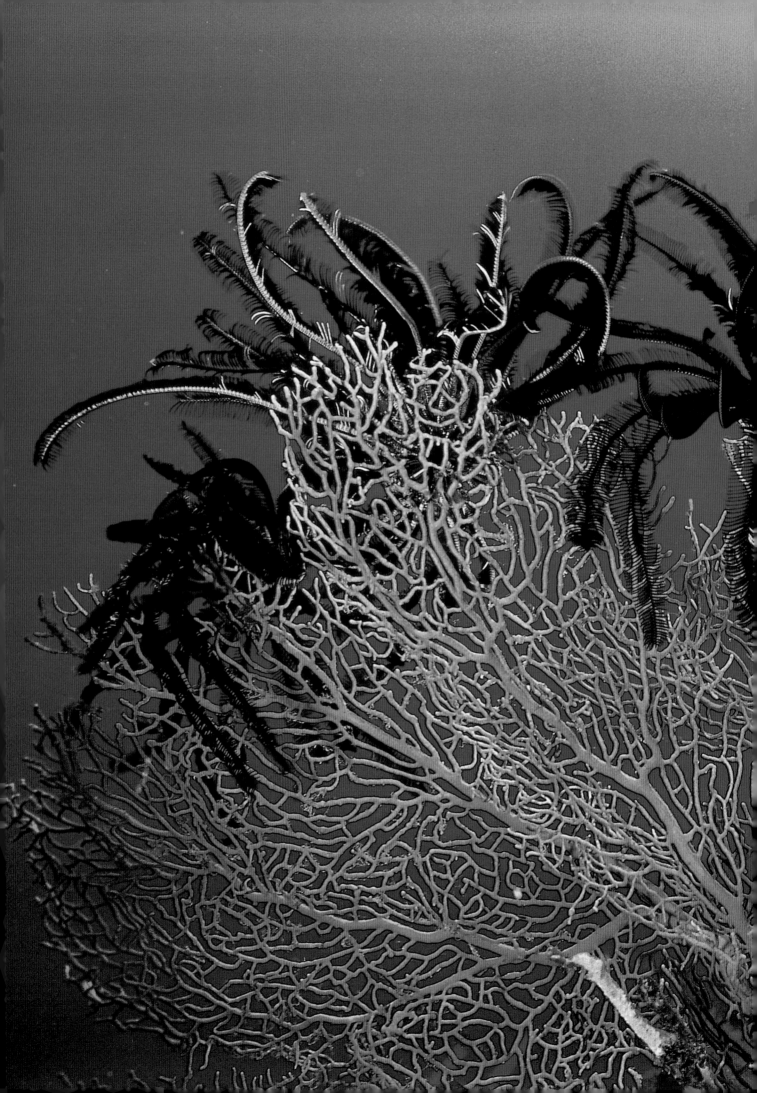

# The Red Sea

**Texts**
Angelo Mojetta

**Translations**
Antony Shugaar

# Contents

INTRODUCTION                                           page    16

- SHARKS, RAYS, MANTA RAYS,
  AND EAGLE RAYS:
  THE GREAT CARTILAGINOUS FISH                         page    24

- BARRACUDAS AND CARANGIDS:
  PREDATORS OF HIGH SEAS                               page    24

- BETWEEN DENS AND GROTTOES                            page    34

- THE QUEENS OF THE SEA BED:
  GROUPERS                                             page    38

- THE FISH OF THE NIGHT                                page    46

- FASCINATING, BUT DANGEROUS                           page    52

- SCHOOLFISH                                           page    60

- THE BUTTERFLIES OF THE REEF                          page    64

- SMALL, CAMOUFLEGED,
  AND ODDLY BEHAVED                                    page    72

- ENORMOUS VARIETY                                     page    78

- PARROTS OF THE SEA                                   page    82

- SURGEONFISH                                          page    86

- TRIGGERFISH AND OTHER ODDITIES                       page    88

- SHELLS,
  PRAWNS AND STARFISH                                  page    94

- THE REEF BUILDERS                                    page   102

14 Blackside
hawkfish
(Parcirrhites forsteri).

# INTRODUCTION

## AN OCEAN CALLED THE RED SEA

Set between the African and Arabian tectonic plates, the Red Sea is the result of their slow but inexorable separation, triggered by the formation of a new sea floor, as a result of the continuous flow of molten material along the ridge that runs down the middle of this sea. For this reason, the Red Sea-aside from being a branch of the Indian Ocean, is considered by geologists to be a genuine, full-fledged ocean in formation. It is expected that, if the phenomena now underway continue with the same intensity, in about one hundred fifty million years, the Red Sea could become as vast as the Atlantic Ocean. The depression of the Red Sea extends more than 2,250 kilometers but only a little more than 1,900 kilometers, of its length is covered with water. To the north, it is blocked off by the Sinai Peninsula, which separates the gulfs of Suez and Aqaba. The Gulf of Suez, which is relatively shallow (about 90 meters at its deepest) and has a relatively flat floor, is about 250kilometers in length, and is 32 kilometers across. The Gulf of Aqaba, 150 kilometers in length, and quite narrow, 16 kilometers across is instead far deeper, with two depressions to the north and south, respectively 1,100 and 1,420 meters in depth. The deepest part of the Gulf of Aqaba is found along the eastern coastline, and is more than 1,800 meters. This measurement, if taken in absolute terms, seems even more impressive since the coastline in this area is surrounded by mountains that rise to elevations of over 1,500 meters. The Red Sea itself, widens out to reach a maximum width of 354 kilometers on a line with Massawa. From there it narrows gradually southward, until it is just 26 kilometers at the Strait of Bab el-Mandab, which then feeds into the Indian Ocean. As for a categorization of the sea floors in the Red Sea, it is possible to distinguish three separate physiographic regions: a coastal region that is as much as 500 meters in depth; an intermediate region of exceedingly irregular configuration, whose lower limit can be set at approximately 1,100 meters; and a deeper region running along the central axis of the Red Sea, where depths can reach as far as 1,500 to 2,000 meters. Of these three ranges, certainly the most interesting for divers is the first and shallowest. In this range, on sea

*16  Glassfish* (Parapriacanthus guentheri) *among soft corals* (Dendronephthya sp.).

*17 Gorgonians and soft corals belonging to various genera* (Dendronephthya, Xenia, Heteroxenia) *settle in the colonies of dead madrepores.*

floors ranging from the surface to 50 meters in depth (just over a fifth of the total area), one will find the coral reefs, more varied and abundant in the central and northern sections of the Red Sea. Because of their characteristics, they can be considered to be a separate whole. The coral reefs found along the coastal range, both to the east and west, belong to the category of fringing reefs, and range in width from a few meters to about a kilometer. Their summits often break the waves at low tide. These coral formations develop atop a rocky sea floor or very compact detritus of organic origin, and they tend to develop lengthwise, out toward the open waters, where the better environmental conditions permit more unhindered growth of the coral polyps. This in turn leads to a progressive expansion of the reef toward the open water, and to the formation of broad channels running parallel to the shoreline; in some cases, and especially in the southern section of the sea and along the Sudanese coasts, these channels can become so broad that they begin to resemble lagoons. The fringing reef, finally, presents another distinctive characteristic.

As one can see from an aerial photograph or from a detailed nautical chart, the reefs are broken occasionally — generally in correspondence with valleys that open out in the mainland — by narrow channels that are extensions of the beds of ancient rivers or the course of temporary rivers that form during the rainy seasons. Depending on the location, these channels, which hover in delicate equilibrium between the growth and the erosion of the coral, take the name of *marsa* or *sharm*. The fringing reefs, however, are not the only coral formations in the Red Sea. Off the coasts of Egypt, Sudan, and Saudi Arabia, on a line with ancient rocky sea floors, which had once emerged from the water because of a decline in the sea level, there are colonies of coral whose continual growth has kept steady pace with the rising sea level, resulting in the appearance of both coral islands and true barrier reefs, several kilometers in length, and surrounded by sea floors that drop sharply away to depths ranging from 500 to 800 meters. Despite the fact that the presence of corals is certainly one of the greatest attractions of this sea, they should still be considered

— at least in certain areas — as an anomaly. The Red Sea, in fact, extends far to the north (as far as 30° N. latitude), and into latitudes where the climatic conditions usually prevent the growth of these organisms, capable of surviving only in waters where the temperature never drops below 10° C. The Red Sea, on the other hand, being a closed body of water, is unusually hot, and there, are minimum temperatures (in February) of around 20-22° C in the northern section and around 27° C in the southern and central basin. The peak temperatures, on the other hand, oscillate around 30-32° C, although in coastal areas with very shallow sea beds and enclosed bays, the temperature can rise as high as 45° C. As for the salinity, the scanty rainfall (just 180 millimeters a year in the area considered rainiest: the region of Suakin), and the very high atmospheric temperatures, which increase evaporation, and the limited water-exchange with the neighboring seas, all help to raize these levels to 38-40 per thousand (i.e., 38-40 grams of salt in every liter of water) with high points that can go to as much as 45-50 per thousand.

## STRUCTURE AND MAKE-UP OF CORAL REEFS

*18 Ras Mohammad, Sinai's southernmost cape.*

*19 Soft corals (Dendronephthya sp.).*

Coral reefs are immense formations, as hard as rock; only the external section of the reef is actually alive. With a very slow rate of growth, on average about a centimeter per year, though of course there are exceptions, these formations are the result of the incessant growth of living organisms, generally known as corals, even though the technical term would be madrepores. Constituted by a living part equipped with numerous tentacles, the polyp, and by a limestone cup-shaped section, the corallite, the corals with which a scuba diver comes into contact during his time underwater are organisms that live in colonies that may vary considerably in shape and size. They are distinguished by a rigid skeleton made up of calcium carbonate and other mineral salts drawn from the sea water through the specific metabolism of these animals. In the Red Sea, specialists have identified more than 170 different species of corals; clearly it is impossible to condense a complete description of them into a few pages. The most common genera are the Acropora and *Montipora*, to which we might well add the *Fungia, Porites, Favia Favites, Stylophora, Pavona, Leptoseris,* and *Cycloseris.* The distribution of the various species is mainly regulated by the environmental conditions (transparency, depths, and hydrodynamics of the water), so that it soon becomes possible to recognize, within certain limits, the existence of preferred areas for one or the other type of coral formation and the recurrence of certain associations of species. And so, if we imagine a dive beginning near the shore and proceeding toward open waters along the reef, we will first find colonies of *Stylophora* (family *Pocipolloridae*). These are ramified colonies with large corallites. The ramifications, slightly compressed and more-or-less flattened at the tips, are light- brown in color, while the extremities of the branches are pink or purple, and the tips are white. Proceeding toward the open sea, we shall encounter colonies of *Favia, Favites,* and *Porites.* The first two genera mentioned can be recognized by their massive colonies, rounded or columnar, or else encrusting (in deep water) and especially for the orderly appearance of their corallites, which look like the honeycomb cells of a beehive (infact, their name comes from the Latin for "hive," in fact). This is their most distinctive feature, while their colors, ranging from whitish to a more-or-less intense pinkish hue, and size (depending upon the age of the colony, as is the case with all corals) are exceedingly variable. The genus *Porites,* which is

among the most common, is a bellwether of sheltered waters with sandy floors. These are large colonies generally rounded in shape, with minute corallites; they resemble compact masses of rock, and are light brown in color. The outermost and most exposed areas of reef, lastly, are dominated by the genus *Acropora*, whose species (15 in the Red Sea) can be considered to be madrepores par excellence. They generally form ramifying colonies, described as "staghorn corals," which are quite common in the first 10 meters of the reef overlooking the open waters. At greater depths, on the other hand, the *Acropora* tend to expand, forming platform-shaped, or umbrella-shaped colonies, with broad flat surfaces supported by a columnar base. Alongside the hard corals, there are other colonial organisms, such as the gorgonians and soft corals, whose shapes and colors are main attractions to the underwater photographers who dive

along the reefs of the Red Sea in search of startling and captivating pictures. In reference to the former, we might point out that they have a corneous skeleton, hard but elastic, which tends to form perpendicular to the current above rocky sea beds, especially in the more exposed and less well lit sections of the reef. A number of gorgonians, known as sea whips because they do not have the usual tree-like branching shape, resemble long stalks, distinguished by a little curl or burr at the end. In the Red Sea, the *Gorgonacea* are not particularly abundant while, on the contrary, the soft corals, or *Alcyonacea* are quite common. The texture of the latter is quite soft or rubbery to the touch, because the limestone component, though it exists, is limited to thin spicules incorporated in the main supporting tissues. These spicules can be clearly seen in the species of the genus *Dendronephthya*, which are among the most common and the loveliest

in the Red Sea, due to their arborescent, or ramifying, appearance. There translucent and brightly colored *Alcyonacea* (fuchsia, pink, or orange) have their polyps clustered together, not unlike the cauliflower-like inflorescences, supported by branches that develop along a single plane or in many directions. These are carnivorous organisms, just like the gorgonians and the hard corals, and for that reason it is more common to see them in an expanded form during the night, when they double or even triple their size; in the case of *Dendronephthya klunzingeri* they can become more than a meter in height. The soft corals are more common in the upper 15 meters from the surface, where there are more-or-less encrusting species, which at a first glance are reminiscent of large *actinia* with short but numerous tentacles, or formations of moss that at times can cover many square meters of the sea bottom.

# THE FISH OF THE RED SEA

Since the Red Sea is a branch of the Indian Ocean, its fish life naturally displays considerable affinities with the fish found in the neighboring ocean, even though the geological history of this sea, which begins about 40 million years ago, has led to the evolution of many endemic species; the latter constitute about 17 percent of the thousand and more species that inhabit the Red Sea.

The chief cause of the abundance of life on the coral sea beds is the remarkable fertility of this environment, which is as varied and complex as its inhabitants. The reefs of the Red Sea, which are always changing, craggy, and broken by underwater crevices and fissures, dotted with pillar, mushroom, or umbrella formations that mark off the sands or the underwater pastures; these formations provide an infinite array of habitats to sea creatures.

Some find this area to be a perfect haven in which to elude the pursuit of predators; others lurk there in ambush, or hide their eggs or their young, or they establish peculiar inter-species.

relationships, like those between clownfish and the sea anemones, the cardinal fish (*Paramia sp., Apogon sp.*) and the black sea urchins, or to name another still, that between the gobies and the crayfish of the genus Alpheus, the two creatures comfortably sharing the same underwater den. All of these reef-dwelling creatures find food there, be that constituted by other fish, plankton, algae, coral polyps, detritus, or other forms of nourishment.

The coral reef varies in more aspects than just the shapes and the seascape. There are also gradients of differentiation, linked to the depth, with such varying factors as hydrodynamics, luminosity, and temperature. Each of these parameters, in the final analysis, affects others still, in accordance with mechanisms whose ultimate product is the abundance of life mentioned here so frequently.

As a result, in just a few ten square meters of reef, one might find hundreds of fish, belonging to dozens of different species, and yet living in perfect harmony, to mention the numerous invertebrates. Only a great number of dives can make a diver familiar with the many different sets of relationships between fish and environment. Inevitably, however, the clearest set of relationships will be those linked to the variety of colors displayed. Even though the colors of the fishes are so exceedingly evident to the eyes of humans, they are not equally evident to the eyes of their fellow fish and even, in many cases, serve to conceal or to camouflage them.

For example, the common ocellate blotches found near the tail or on the dorsal fin, and the dark bands that often cover the eyes, are there to confuse attacking predators, who take the blotches for the real eyes, often attacking the intended prey in sections of the body that are not as vital as expected.

Likewise, the garish colorings made up of blotches or stripes of

*20 Sea fan gorgonian* (Subergorgia hicksoni).

*21 Humphead wrasse* (Cheilinus undulatus).

*22-23 Grey reef shark* (Carcharhinus amblyrhynchos).

contrasting colors, which appear to human eyes as if they were neon advertising signs, actually serve to confuse the fish's outline, breaking it up into many different tiles, or to blend it in with the welter of coral branches or in the lacy patchwork structure of the gorgonians.

But the colorings are not only deceptive in function.

Indeed, in some cases they are used to transmit very specific messages. The groupers, for example, can rapidly change their coloring, accordingly with whether they feel threatened, fearful, or are sleeping and therefore feel tranquil and calm. Indeed, marine biologists have coined the term "pajama coloring" to indicate the markings that certain fishes take on during their nightly repose; these nocturnal markings can be so different from the diurnal colorings that one may think that these are two different species.

In other cases, especially among the Labrids or the parrotfish, the coloring makes it fairly easy to recognize specimens of one gender or the other, or the degree of maturity. And in many species of fish, the difference in color between adults and the young prevents useless battles between species. Lastly, there are colorings that transmit explicit messages, such as: "I am ready for mating," "I am dangerous," or "I am completely inedible."

Equally varied are the forms that the many different fish can take on. Before the eyes of the diver, float, dart, and hover fish of serpentine shape (morays, needle fish, cornet fish), tapered form (groupers, wrasse, tuna, and jacks), globular (globe fish) or else flattened or compressed (angelfish, butterfly fish, surgeon fish, trigger fish) or depressed (rays, eagle-rays, manta rays). To a careful observer, each different shape is clearly linked to the habits of each species and the type of environment in which it lives. This is shown by the fact that coral reefs dominated by a few species of corals, therefore

with little variation, are home to a great number of fish belonging to relatively few species.

The various colonies of corals often house specific populations of small fish. The banded dascyllus (*Dascyllus aruanus*) establishes its territory among the branches of the *Pocillopora* and the *Acropora*.

The latter are also home to populations of blue-green chromis (*Chromis caerulea*) or half-and-half chromis (*C. dimidiata*) along with young specimens of jacks, trigger fish, and boxfish. Of particular interest to an underwater photographer is the umbrella-shaped formation of the *Acropora*; in their shelter one can find angelfish, butterflyfish, surgeonfish, and grunts, while the sands at the base of these formations are often home to blue-spotted lagoon rays (*Taeniura lymma*).

The part of the coral reef that is most abundant in life, however, remains the outer area, which is also frequented by larger fishes, such as sharks.

About thirty different species of shark live in the Red Sea, but the two most common are the white-tip reef shark (*Triaenodon obesus*) and the black-tip reef shark (*Carcharhinus melanopterus*).

In deeper waters, ranging from 10 to 50 meters it is also possible to encounter *Carcharhinus albimarginatus, C. plumbeus* or *C. wheeleri,* while in grottoes and under some coral formations it is possible to find nurse shark (*Nebrius ferrugineus*).

Rare, but far more dangerous, are the tiger shark (*Galeocerdo cuvieri*), the mako (*Isurus glaucus*), and the hammerhead shark (*Sphyrna sp.*).

Fascinating and harmless are the eagle ray (*Aetobatus narinari*), their backs spotted with white, the manta ray (*Manta birostris*) which can attain a width of 5 meters and the devilfish (*Mobula diabolus*).

Among the larger predators that appear suddenly from the open sea we should mention the barracuda (*Sphyraena barracuda,*

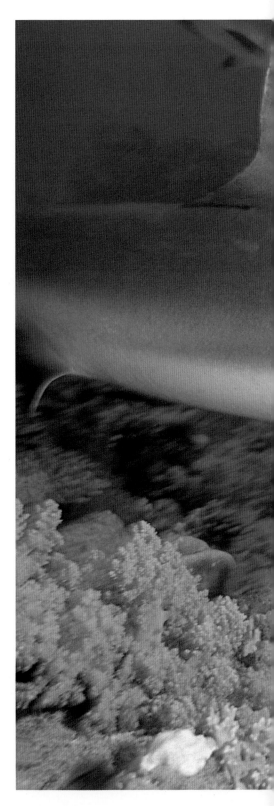

*Sphyraena qenie)* jacks *(Caranx melampygus, C. sexfasciatus, C. ignobilis)*, easy to recognize because of the deeply cleft, forked tail, and the sharply marked lateral line. Mingling with them are often large humphead wrasse (*Cheilinus undulatus*), parrotfish, unicornfish, and swarms of colorful fusiliers (*Caesio sp.*).

Although a dive in tropical seas is a fascinating experience one

should not forget that a coral reef conceals dangers that can be harmful and even fatal, especially if one is ignorant of them.

A number of dangerous or potentially dangerous creatures, such as sharks, barracuda, morays, and scorpionfish, are well known, but they are not the only ones.

A more detailed, and very useful, list, should include the stonefish, both true *(Synanceia verrucosa)* and false *(Scorpaenopsis diabolus)*, dangerous because of their potentially venomous spines, and especially because of their camouflaged coloration, which makes them hard to see.

Equally dangerous though little known are the so-called fire corals *(Millepora dichotoma, M. platyphylla)* which are equipped with powerful stinging cells capable of producing "burns" which should receive immediate care, since they can easily become infected; the mollusks of the genus *Conus*, capable of launching small, poisonous darts; sea urchins, both the long-spined *Diadema* and the *Astenosoma*, whose bright red or green coloring may well attract a diver—caution is the watchword, because the short white- and red-tipped spines are quite venomous.

# *S*HARKS, RAYS, MANTA RAYS, AND EAGLE RAYS: THE GREAT CARTILAGINOUS FISH

# *B*ARRACUDAS AND CARANGIDS: PREDATORS OF HIGH SEAS

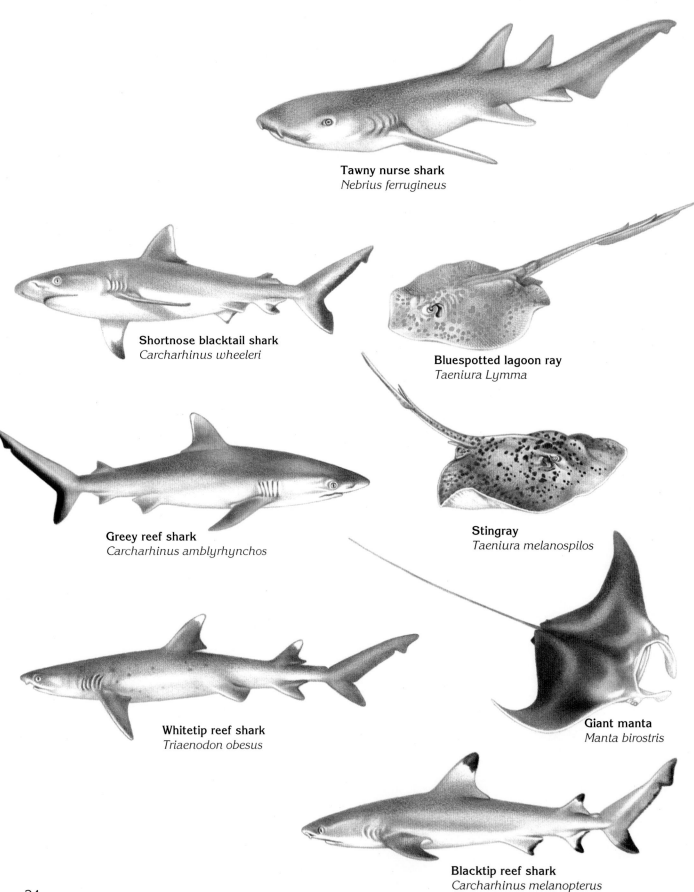

**Tawny nurse shark**
*Nebrius ferrugineus*

**Shortnose blacktail shark**
*Carcharhinus wheeleri*

**Bluespotted lagoon ray**
*Taeniura Lymma*

**Greey reef shark**
*Carcharhinus amblyrhynchos*

**Stingray**
*Taeniura melanospilos*

**Whitetip reef shark**
*Triaenodon obesus*

**Giant manta**
*Manta birostris*

**Blacktip reef shark**
*Carcharhinus melanopterus*

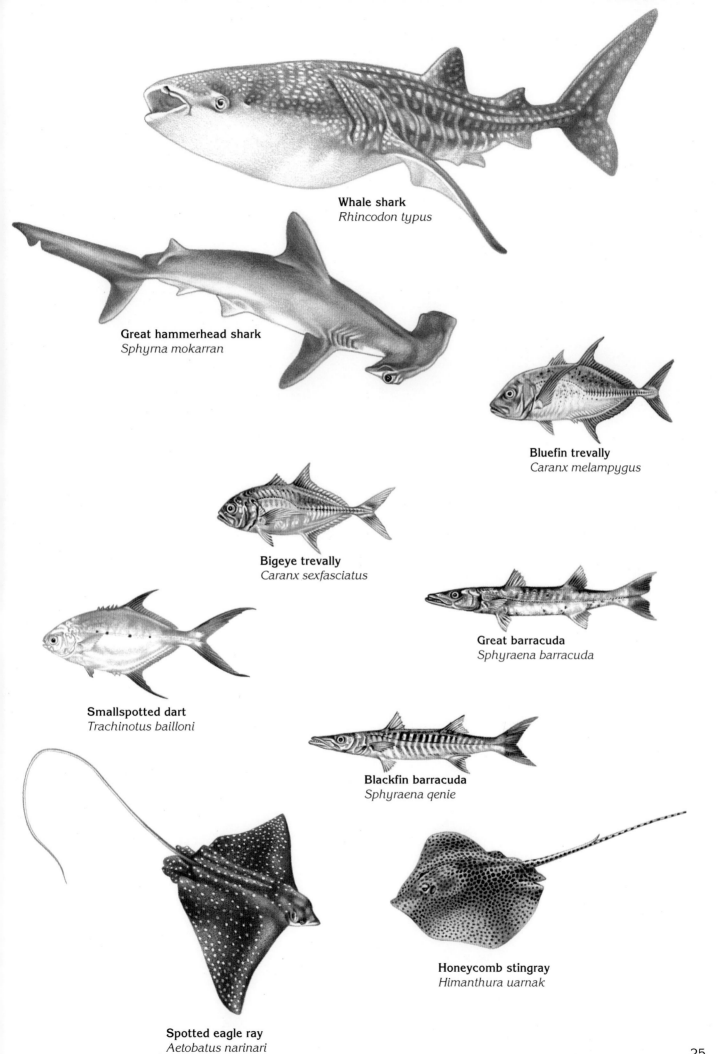

**Whale shark**
*Rhincodon typus*

**Great hammerhead shark**
*Sphyrna mokarran*

**Bluefin trevally**
*Caranx melampygus*

**Bigeye trevally**
*Caranx sexfasciatus*

**Great barracuda**
*Sphyraena barracuda*

**Smallspotted dart**
*Trachinotus bailloni*

**Blackfin barracuda**
*Sphyraena qenie*

**Honeycomb stingray**
*Himanthura uarnak*

**Spotted eagle ray**
*Aetobatus narinari*

25

26-27
*Whale shark*
(Rhincodon typus).

Intriguing and far less dangerous than is still commonly believed, sharks are frequent visitors to the reefs of the Red Sea. Endowed with a cartilaginous skeleton, and not a bony one like those of the other fish, and without a swimming bladder—which means that many sharks are forced to swim incessantly in order to maintain the desired depth—sharks are tapered and hydrodynamic in shape, made particularly distinctive by the tailfin with the larger upper lobe. The shortnose blacktail shark and the blacktip reef shark (*Carcharhinus wheeleri, C. melanopterus*), the whitetip reef shark (*Triaenodon obesus*), and the tawny nurse shark (*Nebrius ferrugineus*) are among the sharks most frequently encountered along the coral reefs, but while the first two swim tirelessly, appearing and vanishing from the view of the scuba divers, the whitetip reef shark and the tawny nurse shark are easy to watch, as they lie resting during the day in their underwater grottoes. Rarer, but not impossible to spot, are the oceanic whitetip sharks (*C. longimanus*), the hammerhead shark (*Sphyrna sp.*), and the whale shark (*Rhincodon typus*), which is gigantic (more than 12 meters in length), but entirely harmless.

*26 bottom*
*Great hammerhead*
(Sphyrna
mokarran).

*27 Tawny nurse*
*shark* (Nebrius
ferrugineus).

*28 top*
*Whitetip reef shark*
(Triaenodon
obesus).

*28 center*
*Oceanic whitetip*
*shark*
(Carcharhinus
longimanus).

*28 bottom and 28-29*
*Shortnose blacktail*
*shark* (Carcharhinus
wheeleri).

Cartilaginous fish like the sharks, but entirely different in shape due to an evolutionary history that led them to live on the ocean floor, are the great rays (the bluespotted lagoon ray and the Honeycomb stingray, or *Taeniura lymma* and *Himanthura uarnak*), often found on sandy expanses where they lie poised like airplanes ready for takeoff. Belonging to the same order (Raiformes) are the spotted eaglerays (*Aetobatus narinari*), with their great pointy pectoral fins and long tails, and the enormous, majestic giant mantas (*Manta birostris*), black and white in color, with fins that attain a width of 5 to 6 meters, marked by their two cephalic fins shaped like mitts, which they use to convey food toward their mouths.

*30-31*
*Spotted eagle ray*
(Aetobatus
narinari).

*30 bottom*
*Bluespotted*
*lagoon ray*
(Taeniura lymma).

*31 top*
*School of spotted*
*eagle ray*
(Aetobatus
narinari).

*31 bottom*
*Giant manta*
(Manta birostris).

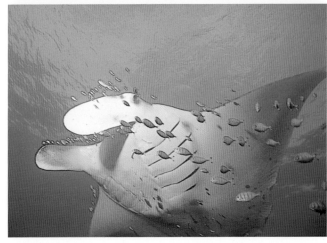

Sharing their habits as predatory fish that live along the boundary between the reef and the open sea, barracudas (*Sphyraena sp.*) and carangids (the bluefin trevally and the bigeye trevally, or *Caranx melampygus* and *C. sexfasciatus*) move in schools that can number dozens and dozens of specimens, in search of their habitual prey, which include the smaller schoolfish; they attack these smaller fish after herding them toward the reef so as to hinder their escape. Barracudas are considered to be potentially dangerous to humans, but in reality the very few attacks on record, with expert verification, have proven to occur in murky waters, and to have involved particularly large and solitary barracudas, or else barracudas that were attacked or disturbed by a scuba diver.

*32 top*
*Yellowspotted jack*
(Carangoides bajad).

*32 center*
*Silver Pompano*
(Trachinotus blochii).

*32 bottom*
*Bigeye jack*
(Caranx sexfascitus).

*33 top*
*Bluefin trevally*
(Caranx melampygus).

*32 -33*
*Blackfin barracuda*
(Sphyraena qenie).

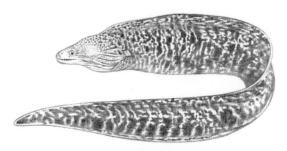

**Undulated moray**
*Gymnothorax undulatus*

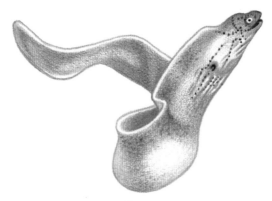

**Grey moray**
*Siderea grisea*

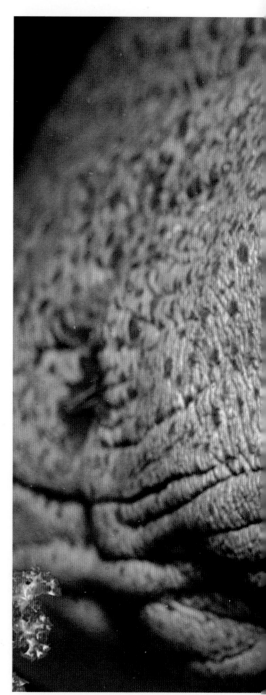

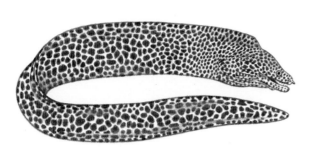

**Honeycomb moray**
*Gymnothorax favagineus*

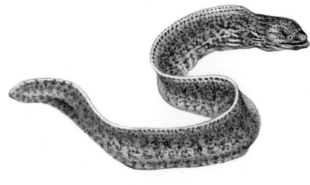

**Giant moray**
*Gymnothorax javanicus*

**Cornetfish**
*Fistularia commersonii*

**Shark sucker or remora**
*Echeneis naucrates*

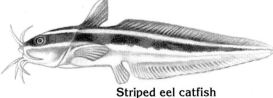

**Striped eel catfish**
*Plotosus lineatus*

*34-35*
*Giant moray*
(Gymnothorax
javanicus).

Elongated in shape, sinuous, and similar to snakes —
except for the remora — the fish described on these
pages are typical sea bed dwellers and feeders.
The morays (e.g., the grey moray and the giant moray,
respectively *Siderea grisea* and *Gymnothoraxjauanicus*)
are commonly encountered amidst the crannies of the
reefs. There are numerous underwater grottoes and
crannies that these animals choose as their refuge
during the day; from the entrances of these cavities,
they extend their open mouths showing their powerful
front teeth in a display that seems threatening but, in
fact, is not. The morays keep their mouths agape
because this is their way of breathing.

*36 top*
*Grey moray*
(Siderea grisea).

*36 bottom*
*Young hundulated*
*moray*
(Gymnothorax
undulatus).

*37 top left*
*Spotted garden eel*
(Heteroconger
hassi).

*37 top right*
*Remora*
(Echeneis
naucrates).

*37 bottom*
*Cornetfish*
(Fistularia
commersonii).

Much cannier and more cunning are the so-called garden eels, a term which refers to their odd life-style. These fish (*Heteroconger hassi*) live in very sizable groups on the sandy sea floors, where each one digs its own den, with part of their bodies protruding, though they never entirely leave the little tunnel that they have dug in the sand.

From a distance, it is possible to see dozens and dozens of these creatures undulating in the water, but one need only get a few meters closer, and the eels will vanish in a flash.

Elongated in shape, but with a narrow tubular snout that ends in a pair of thick lips, the cornetfish

(*Fistularia commersonii*) move slowly and rigidly through the water, counting on their innocuous appearance and especially on their uniform coloring, to venture quite close to smaller fish, who are then literally sucked into the cornetfish's large mouth. Perfectly evolved to hitch rides upon larger fish, and therefore exceedingly difficult to spot on their own, remoras (*Echeneis naucrates*) are constant companions of sharks and mantas, to whose bodies they fasten by virtue of having the anterior dorsal fin converted into an oval transversely lamellate suctorial disc on the top of the head, by means of which they adhere firmly.

# THE QUEENS OF THE SEA BED: GROUPERS

39  Giant grouper (Epinephelus tauvina).

**Coral grouper**
*Cephalopholis miniata*

**Potato cod**
*Epinephelus tukala*

**Peacock grouper**
*Cephalopholis argus*

**Olive dottyback**
*Pseudochromis fridmani*

**Redmouth grouper**
*Aethaloperca rogaa*

**Scalefin anthias**
*Pseudanthias squamipinnis*

**Smalltooth grouper**
*Epinephelus microdon*

**Sunrise dottyback**
*Pseudochromis flavivertex*

**Giant grouper**
*Epinephelus tauvina*

**Lunartail grouper**
*Variola louti*

Among the fish most commonly found on coral reefs, ranging from depths of just a few meters to well beyond the limits to which a scuba diver can safely venture, one will certainly find groupers, almost all of them distinctive for their lively colors, powerful bodies, broad tailfins, and large mouths, with the lower jaw larger than the upper. The considerable variety of environments that one can encounter during a dive along the coral reefs of the Red Sea explains the great number of species of groupers that one can observe. There are small colorful groupers, such as the coral grouper (*Cephalopholis miniata*) and the peacock grouper (*C. argus*), there are mid-sized groupers such as the lunartail grouper (*Variola louti*), distinguished by a caudal fin shaped like a crescent or sickle, that is unique among the groupers, and the redmouth grouper (*Aethaloperca rogaa*) all the way up to the giants of the group, such as *Epinephelus tukula*, the potato cod, which can grow to be as long as two meters. Although some species are more frequent by day, others at dawn or sunset, and still others by night, groupers all have territorial habits (each one lives in a den where it spends much of its time and where it takes refuge if it is disturbed) and are all carnivorous, feeding on fish, crustaceans, and octopuses.

*40-41*
*Coral grouper*
(Cephalopholis
miniata).

*41 top*
*Giant grouper*
(Epinephelus
tauvina).

*41 bottom*
*Potato cod*
(Epinephelus
tukula).

42 top
*Lunartail grouper*
(Variola louti).

42-43
*Peacock grouper*
(Cephalopholis
argus).

43 top
*Red Sea grouper*
(Plectropomus p.
marisrubri).

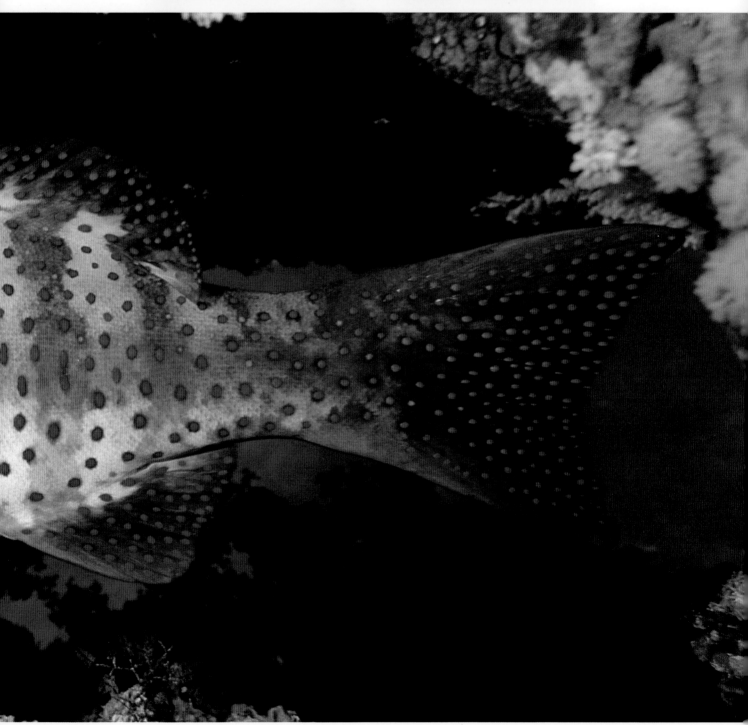

The orange clouds of color and life that surround the more extensive and jagged coral formations, on the other hand, are made up of little fish belonging to the same family as the grouper, the *Serranidae*.
These little fish are scalefin anthias (*Pseudanthias squamipinnis*), which attain lengths of up to 15 centimeters; they live in isolated groups, each one near a coral formation. Oddly enough, each group could more accurately be described as a harem, inasmuch as there is a dominant male surrounded by females.

The males are distinguished by their brighter coloring, but especially by the long ray set in the first section of the anterior fin. Since these fish change sex as they mature, upon the death of the male, the largest female assumes the dominant role. The coral nooks and fissures are home to other small fish with a shape more tapered than the scalefin anthias; these splendid fish are spectacularly colorful, such as the olive dottyback, or *Pseudochromis fridmani*, with its fluorescent violet hue, which is found only in the Red Sea.

*45 top*
*Jewel fairy basslet*
*male*
(Pseudanthias
squamipinnis).

*45 center*
*Jewel fairy basslet*
*female*
(Pseudanthias
squamipinnis).

*45 bottom*
*Olive dottyback*
(Pseudochromis
fridmani).

*44-45  Scalefin*
*anthias female*
(Pseudanthias
squamipinnis).

# *T*HE FISH OF THE NIGHT

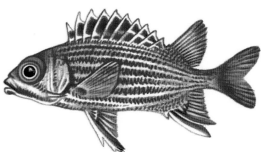

**Crown squirrelfish**
*Sargocentron diadema*

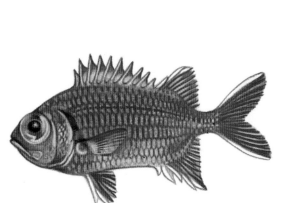

**Blotcheye soldierfish**
*Myripristis murdjan*

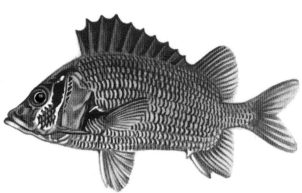

**Sabre squirrelfish**
*Sargocentron spiniferum*

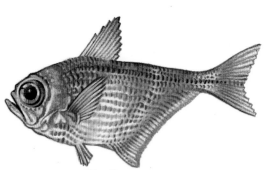

**Vanikoro sweeper**
*Pempheris vanicolensis*

**Golden cardinalfish**
*Apogon aureus*

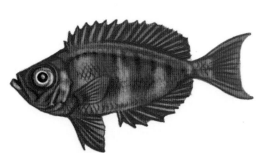

**Goggle eye**
*Priacanthus hamrur*

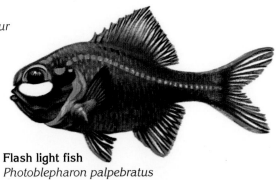

**Flash light fish**
*Photoblepharon palpebratus*

47
*Sabre squirrelfish*
(Sargocentron
spiniferum).

At the mouths of underwater grottoes or inside those grottoes, it is possible to observe a number of fish during the day that are linked by their coloring, based on red, and by their large eyes. They are not very active, and lie practically motionless under the gaze of the scuba diver, giving the impression that they are imprisoned in a glass display case. In reality, these fish (squirrelfish, soldierfish, cardinalfish, and *Priacanthidae* belong, respectively, to the genera *Sargocentron, Myripristis, Apogon,* and *Priacanthus*) are nocturnal and evening species, who dislike bright light and therefore take refuge in the underwater grottoes or in the less brightly lit areas; they emerge at nightfall to take the place of the diurnal fish, so that life on the reef never slows down.

*48 top and 48-49*
*Coggle eye*
*(Priacanthus*
*hamrur).*

*48 bottom*
*Sabre squirrelfish*
*(Sargocentron*
*spiniferum).*

48

50-51
*Glass fish*
(Parapriacanthus
guentheri).

*50 bottom*
*Young lyretail*
*hogfish*
(Bodianus
anthioides).

Equally fearful of light are the *Pempheridae*, small fish (*Pempheris vanicolensis*) that loiter in large schools in the shadier areas, where they remain until sunset, when they scatter away across the reef, just like the other fish described here, in search of food. These are commonly known to scuba divers as "glassfish" because of their glittering coloration, which reflects the light of the photographer's flashgun like a many-faceted mirror.

*51 top*
*Vanikoro sweeper*
(Pempheris
vanicolensis)

*51 bottom*
*Glass fish*
(Parapriacanthus
guentheri).

# *Fascinating, but dangerous*

**Frogfish**
*Antennarius coccineus*

**Clearfin turkeyfish**
*Pterois radiata*

**Turkey fish**
*Pterois volitans*

**Devil scorpionfish**
*Scorpaenopsis diabolus*

**Stonefish**
*Synanceia verrucosa*

**Lizardfish**
*Synodus variegatus*

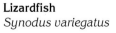

**Pixy hawkfish**
*Cirrhitichthys oxycephalus*

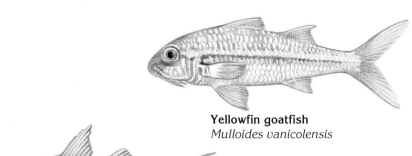

**Yellowfin goatfish**
*Mulloides vanicolensis*

**Yellow saddle goatfish**
*Parupeneus cyclostomus*

**Forsskal goatfish**
*Parupeneus forsskali*

*52-53  Turkey fish*
(Pterois volitans).

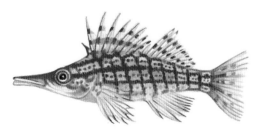

**Longnose hawkfish**
*Oxycirrhites typus*

**Crocodile fish**
*Cociella crocodila*

53

Keeping a careful eye on the sea floor over which one is swimming and on the waters around, allows alert scuba divers in the Red Sea to avoid some of the greatest menaces to which they are exposed. These dangers are the stonefish (*Synanceia verrucosa*), truly quite similar in appearance to little pieces of rock; they can only be distinguished from actual stones by a sharp eye, capable of making out the nearly vertical profile of the mouth and the moving eyes. For that reason, it is wise to be particularly careful of how one places one's hands on the coral floors. The chief threat posed by these fish lies in their very strong spinal rays (they can even penetrate beach shoes!) linked to glands that secrete a highly toxic venom comparable to that of the most lethally venomous snakes. Equally camouflaged, but less dangerous,

*54  Clearfin turkeyfish (Pterois radiata).*

*55 top Stonefish (Synanceia verrucosa).*

*55 center Scorpionfish (Scorpaenopsis barbatus).*

*55 bottom Filament finned stinger (Inimicus filamentosus).*

are the devil scorpionfish *(Scorpaenopsis diabolus)*, similar in shape and behavior to the scorpionfish of Mediterranean waters. More elegant and lovely, but not less dangerous are the *Scorpaenidae* of the genus *Pterois* (turkeyfish and clearfin turkeyfish, respectively *Pterois volitans* and *P. radiata*), distinguished by their fins with long rays, similar to plumes, which actually conceal venomous spines. If carefully approached, they can be watched in peace, as long as one is certain not to corner them in situations where they might feel hemmed in or in danger, so that they might attack, their venomous spines levelled forward.

*56 top*
*Longnose*
*hawkfish*
(Oxycirrhites
typus).

*56 bottom*
*Goldsaddle*
*goatfish*
(Parupeneus
cyclostomus).

*56-57*
*Blackside*
*hawkfish*
(Paracirrhites
forsteri).

Surrounded by myriads of colorful fish that swim all around him, a scuba diver is likely to lose sight of the sea floor, and with it, the opportunity to observe fish that may be less chromatically attractive than others, but equally remarkable and interesting in their ways of life. Very much like the mullets of the Mediterranean Sea, the yellowsaddle goatfish and Forsskal's goatfish of the Red Sea (*Parupeneus forskali* and *P. cyclostomus*) dart along near the bottom, sampling the sediment with their long and mobile barbles, dense with sensory cells. Smaller and brightly colored, but equally well camouflaged are the longnose hawkfish (*Oxycirrhites typus*) and the blackside hawkfish (*Parracirrhites forsteri*), which tend to perch on the sea floor or on the branches of gorgonians, where the longnose hawkfish in particular blend in by virtue of their coloring made up of white and red checks, and because of their small size (13 centimeters).

58 top
*Spekled sandperh*
(Parapercis
hexophtalma).

*58 bottom*
*Crocodile fish*
(Cociella
crocodila) .

*58-59*
*Lizardfish*
(Synodus
variegatus).

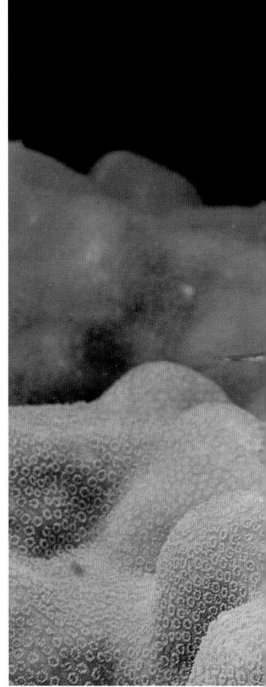

More difficult to see are the lizardfish and the crocodile fish. The former (*Synodus variegatus*) wait in ambush on the sea floor, where their colors camouflage them, making them look like pieces of abandoned coral. The latter (*Cociella crocodila*), instead, with their flattened bodies and long, wide snouts, like to bury themselves among the sediment, leaving only part of their head, their eyes, and their dorsal fins protruding from the dirt.

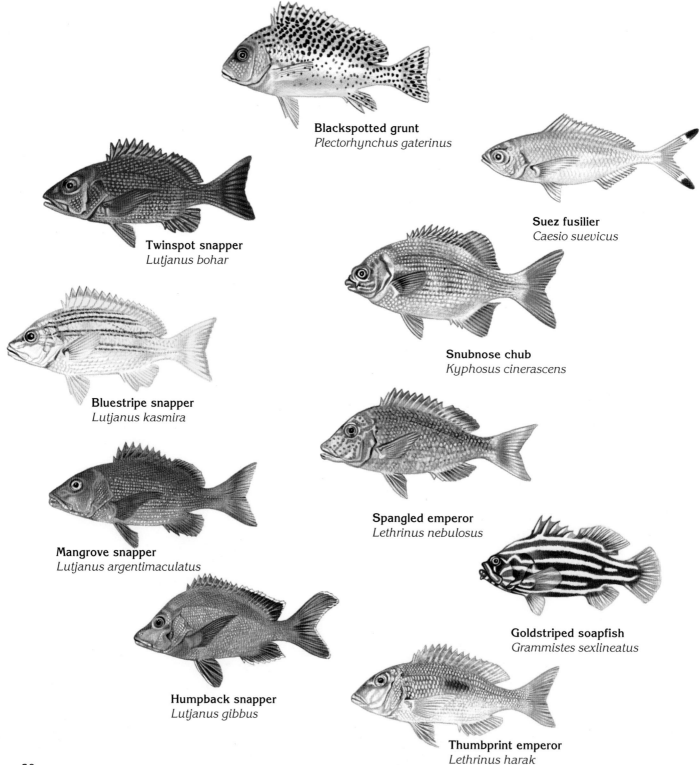

**Blackspotted grunt**
*Plectorhynchus gaterinus*

**Suez fusilier**
*Caesio suevicus*

**Twinspot snapper**
*Lutjanus bohar*

**Snubnose chub**
*Kyphosus cinerascens*

**Bluestripe snapper**
*Lutjanus kasmira*

**Spangled emperor**
*Lethrinus nebulosus*

**Mangrove snapper**
*Lutjanus argentimaculatus*

**Goldstriped soapfish**
*Grammistes sexlineatus*

**Humpback snapper**
*Lutjanus gibbus*

**Thumbprint emperor**
*Lethrinus harak*

These species of fish swimming amongst the corals and amidst the reefs astonish observers with their frequently colorful markings (for example, the fusiliers of the genus Caesio, the Haemulidae, such as the blackspotted grunt or sweetlips, *Plectorhynchus gaterinus*) and with their remarkable size, which can reach lengths of more than 55 or 60 centimeters (twinspot snapper, or *Lutjanus bohar*). In part, predators, and in part, plankton-eaters, whether diurnal or nocturnal, these species indicate by their mere presence the areas richest in food, and are among the first — especially the larger Lutjanidae, called snappers for a good reason to reach areas where the larger predators have just finished fighting over prey, eager to snap up a bit of the leftovers.

*62-63*
*Bluestripe snapper*
(Lutjanus
kasmira).

*63 top*
*Twinspot snapper*
(Lutjanus bohar).

*63 center*
*Spangled emperor*
(Lethrinus
nebulosus).

*63 bottom*
*Lunar fusilier*
(Caesio lunaris).

# The Butterflies of the Reef

**Paleface butterflyfish**
*Chaetodon mesoleucos*

**Orangeface butterflyfish**
*Gonochaetodon larvatus*

**Striped butterflyfish**
*Chaetodon fasciatus*

**Emperor angelfish**
*Pomacantus imperatur*

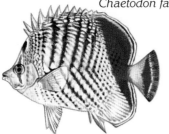

**Crown butterflyfish**
*Chaetodon paucifasciatus*

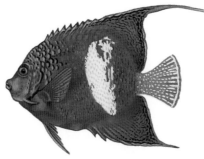

**Yellowbar angelfish**
*Pomacantus maculosus*

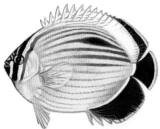

**Exquisite butterflyfish**
*Chaetodon austriacus*

**Arabian angelfish**
*Pomacantus asfur*

**Red Sea bannerfish**
*Heniochus diphreutes*

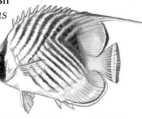

**Threadfin butterflyfish**
*Chaetodon auriga*

**Pennantfish**
*Heniochus intermedius*

**Masked butterflyfish**
*Chaetodon semilarvatus*

**Blackback butterflyfish**
*Chaetodon melannotus*

**Circularbatfish**
*Platax orbicularis*

**Longfin batfish**
*Platax teira*

*65 Batfish*
(Platax
orbicularis).

**Royal angelfish**
*Pygoplites diacanthus*

Without these splendid fish the coral reefs of the tropics would probably be far less fascinating. These fish, in fact, belong to the families Chaetodontidae (butterflyfish), *Pomacanthidae* (angelfish), and *Platacidae* (batfish), which include the major part of the most colorful fish. Bound to the coral reefs and formations, where they find food and shelter, the *Chaetodontidae* and the *Pomacanthidae* have developed laterally flattened shapes that allow them to swim easily even among the most labyrinthine of ramifications.

The variety of coloring, moreover, not only allows the species to recognize each other and to distinguish the young from the sexually mature adults, but is also useful in concealment from other animals.

The stripes, bands, and spots, which appear to the human eye so colossally garish, do not look the same at all to the rest of the underwater kingdom, where the predators of these fish see them as broken-up images, formless or even two-headed. That, for instance, is precisely the function served by those splotches that are known as "ocellar" — hence the name — which can be seen in the rear portion of many butterflyfish and some angelfish. Usually these splotches are associated with dark bands that cover the eyes, making the deception more plausible, leading the predators to attack the tail, mistaken for the head.

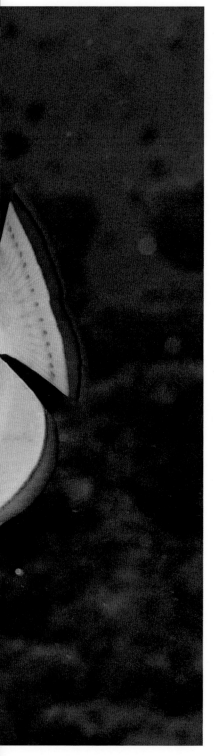

*66-67*
*Striped butterflyfish*
(Chaetodon
fasciatus).

*67 top*
*Threadfin butterflyfish*
(Chaetodon auriga).

*67 center*
*Crown butterflyfish*
(Chaetodon
paucifasciatus).

*67 bottom*
*Lined butterflyfish*
(Chaetodon
lineolatus).

Among the most distinctive of the species found in these families, we should mention the masked butterflyfish (*Chaetodon semilarvatus*) which is often found in schools, the crown butterflyfish (*C. paucifasciatus*), the schooling bannerfish (*Heniochus intermedius*), the Arabian angelfish (*Pomacanthus asfur*), dark blue, with a broad yellow band extending from the belly to the back. Flattened, tall bodies are also found in the batfish (*Platax orbicularis*) which measure as much as 50 or 60 centimeters in length, and nearly the same in height, so that they resemble disks, especially the adults.

*68 top left and 69*
*Royal angelfish*
(Pygoplites
diacanthus).

*68 top right*
*Yellowbar angelfish*
(Pomacanthus
maculosus).

*68 bottom left*
*Emperor angelfish*
(Pomacanthus
imperator).

*68 bottom right*
*Arabian angelfish*
(Pomacanthus
asfur).

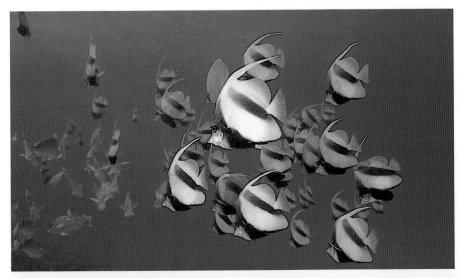

*71 top*
*Red Sea*
*bannerfish*
(Heniochus
diphreutes).

*71 center*
*Masked*
*butterflyfish*
(Chaetodon
semilarvatus).

*71 bottom*
*Masked*
*butterflyfish*
(left) *and*
*schooling*
*bannerfish* (right).

*70 Red Sea*
*bannerfish*
(Heniochus
intermedius).

**Banded dascyllus**
*Dascyllus aruanus*

**Domino damselfish**
*Dascyllus trimaculatus*

**Twobar anemonefish**
*Amphiprion bicinctus*

**Sixspot goby**
*Valenciennea sexguttata*

**Half-and-half chromis**
*Chromis dimidiata*

**Sulphur damselfish**
*Pomacentrus sulfureus*

**Sergeant major**
*Abudefduf saxatilis*

**Mimic blenny**
*Aspidontus taeniatus*

**Sergeant scissortail**
*Abudefduf sexfasciatus*

*73 Whipcoral dwarfgoby (Bryaninops youngei).*

**Blue-green chromis**
*Chromis viridis*

Reproduced in hundreds if not thousands of pictures, the *Pomacentridae* are fundamentally represented by the clownfish (*Amphiprion sp.*) which, due to their characteristics, are considered a group to themselves. Despite the fact that they are so widely known, they still never fail to attract the attention of scuba divers, especially if those divers are also underwater photographers, due to their odd habit of living in close proximity to the stinging sea anemones, which constitute an impregnable defense against would-be predators. Often scuba divers have a chance to observe the female caring for the eggs laid at the base of the anemones, helped by the male who intently wards off intruders with the remarkable aggressivity that so contrasts with the usual behavior of the little clownfish. Along with the *Amphiprion*, one can often observe small black fish with three white spots. These are the young of the domino damselfish (*Dascyllus trimaculatus*), the adults of which,

however, live among the coral formations along with dozens of other specimens belonging to countless species of damselfish, such as the banded dascyllus (*Dascyllus aruanus*), the bluegreen chromis, and the half-and-half chromis (respectively, *Chromis caerulea* and *C. dimidiata*), entirely similar to their counterparts in the Mediterranean. Larger and unmistakable due to their coloring in broad white and black bands are the sergeant major and the sergeant scissortail (respectively, *Abudefduf saxatilis* and *A. sexfasciatus*) which gather in great closely packed groups in the water, surrounding and following scuba divers, without fear. Smaller, and generally found on the sea floor, in the countless nooks and crannies of the corals, are the *Blennidae*, or blennies, many of which have distinctive barbs or fringed filaments on their heads, and the *Gobiidae*, or gobies, with short snouts and ventral fins transformed into suctorial disks with which they adhere to the sea beds.

74-75
Twobar
anemonefish
(Amphiprion
bicinctus).

75 center
Red Sea mimic
blenny
(Ecsenius
gravieri).

75 top
Sergeant major
(Abudefduf
saxatilis).

75 bottom
Lemon goby
(Gobiodon
citrinus).

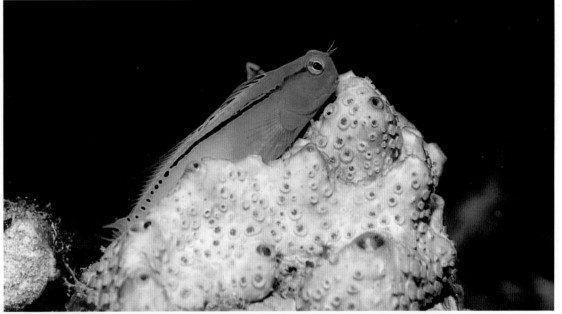

*76 top*
*Banded dascyllus*
(Dascyllus aruanus).

*76 bottom*
*Domino damselfish*
(Dascyllus
trimaculatus).

*76-77  Chromis*
(Chromis sp.).

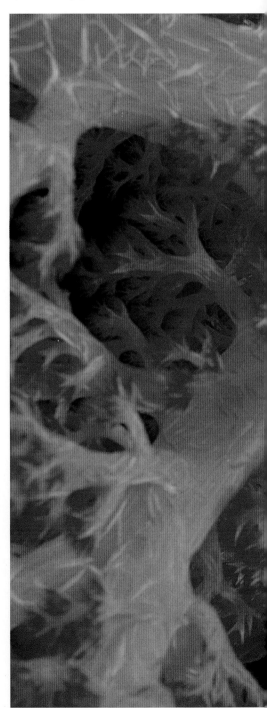

*77 top*
*Blue-green chromis*
(Chromis caerulea).

**Klunziger's wrasse**
*Thalassoma klunzigeri*

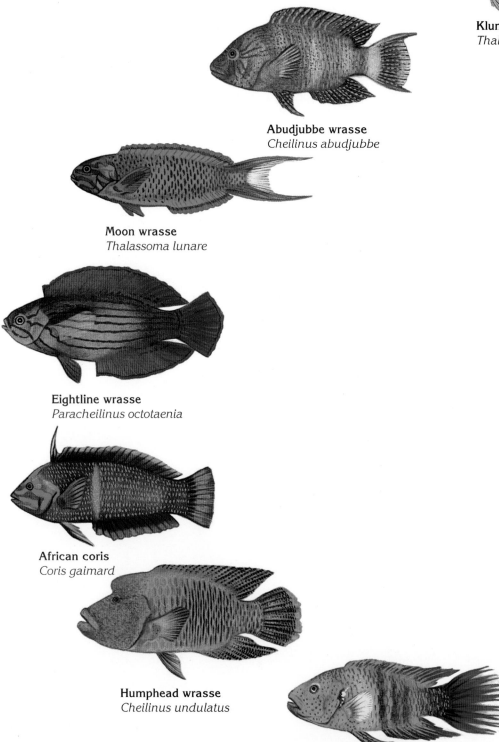

**Abudjubbe wrasse**
*Cheilinus abudjubbe*

**Moon wrasse**
*Thalassoma lunare*

**Eightline wrasse**
*Paracheilinus octotaenia*

**African coris**
*Coris gaimard*

**Humphead wrasse**
*Cheilinus undulatus*

**Broomtail wrasse**
*Cheilinus lunulatus*

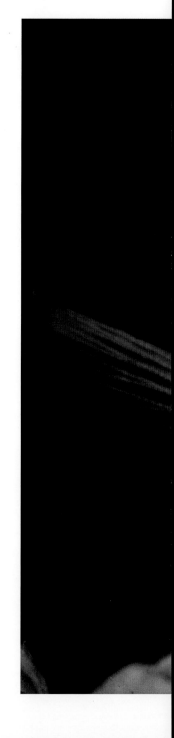

**Red Sea bird wrasse**
*Gomphosus caeruleus*

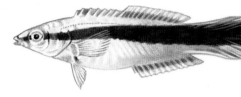

**Cleaner wrasse**
*Labroides dimidiatus*

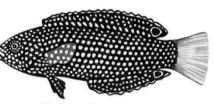

**Yellowtail wrasse**
*Anampses meleagrides*

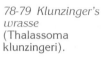

78-79 *Klunzinger's wrasse* (Thalassoma klunzingeri).

**Axilspot hogfish**
*Bodianus axillaris*

Extremely widespread, but differing radically in shape, coloring, and size, the Labridae do resemble each other in their elongated body shape, which is slightly compressed, and in the single long dorsal fin. Their way of swimming is also quite distinctive. The *Labridae*, in fact, swim by propelling themselves forward or backward with powerful strokes of their pectoral fins, which they use like oars; this gives the fish a distinctive undulating movement that one can easily learn to recognize, even at a distance. The most distinctive members, and at the same time fair representatives of the immense variety of the family, are the cleaner wrasse (*Labroides dimidiatus*), with dark-blue and black horizontal bands; these fish are incessantly occupied in cleaning other fish of parasites and organic residue,

and the huge humphead wrasse (*Cheilinus undulatus*) can grow to lengths of up to two meters. All of the *Labridae* are decidedly diurnal, and it is therefore impossible to see them swim at night. With some patience and care, however, it is possible to find them in their night-time hiding places, as they sleep half-buried in the sand or lying on one side in the shelter of a rock or a grotto, or else wrapped in a capsule of mucous, like that used by the parrotfish. Moreover, many of them change color, taking on drabber and less flashy colorings. And for that matter, variations in color are one of the distinctive characteristics of the *Labridae,* and many species change color sharply when, in the process of maturing, they change sex from female to male.

81 top
*Female sling-jaw
wrasse* (Epibulus
insidiator).

81 bottom
*Moon wrasse*
(Thalassoma
lunare).

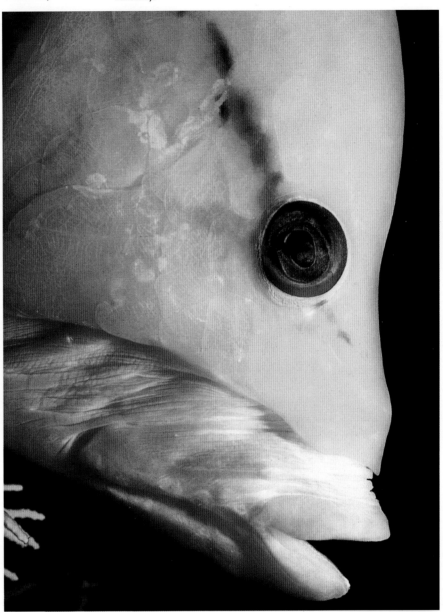

*80-81  Humphead
wrasse* (Cheilinus
undulatus) *and
shark sucker*
(Echeneis
naucrates).

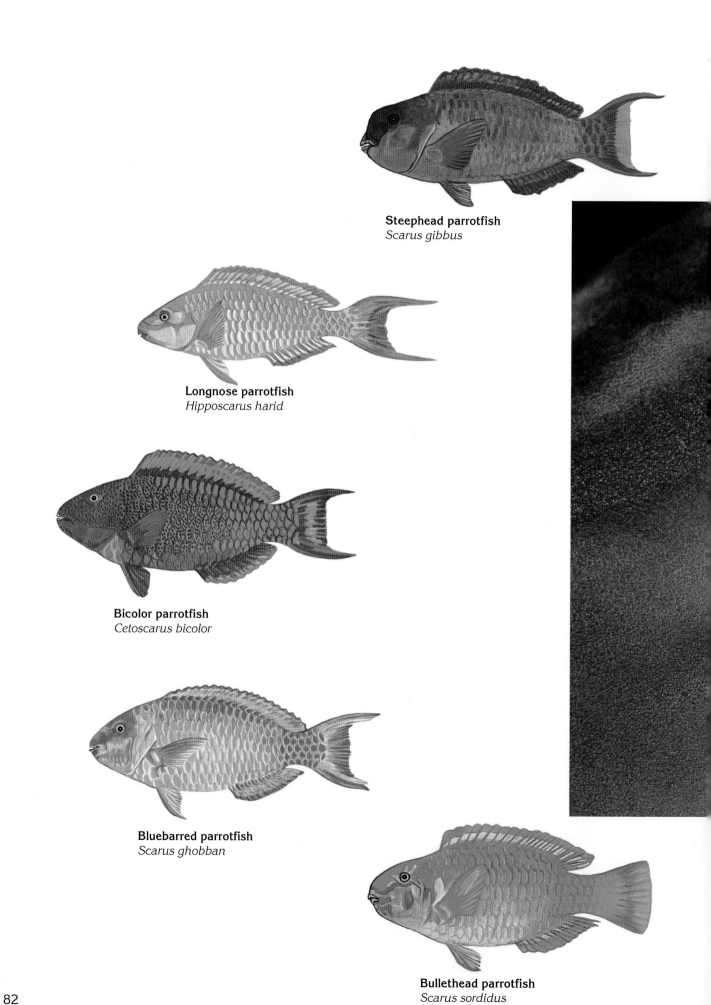

**Steephead parrotfish**
*Scarus gibbus*

**Longnose parrotfish**
*Hipposcarus harid*

**Bicolor parrotfish**
*Cetoscarus bicolor*

**Bluebarred parrotfish**
*Scarus ghobban*

**Bullethead parrotfish**
*Scarus sordidus*

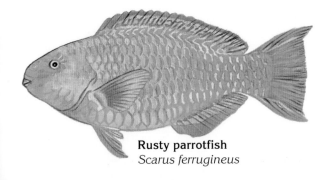

**Rusty parrotfish**
*Scarus ferrugineus*

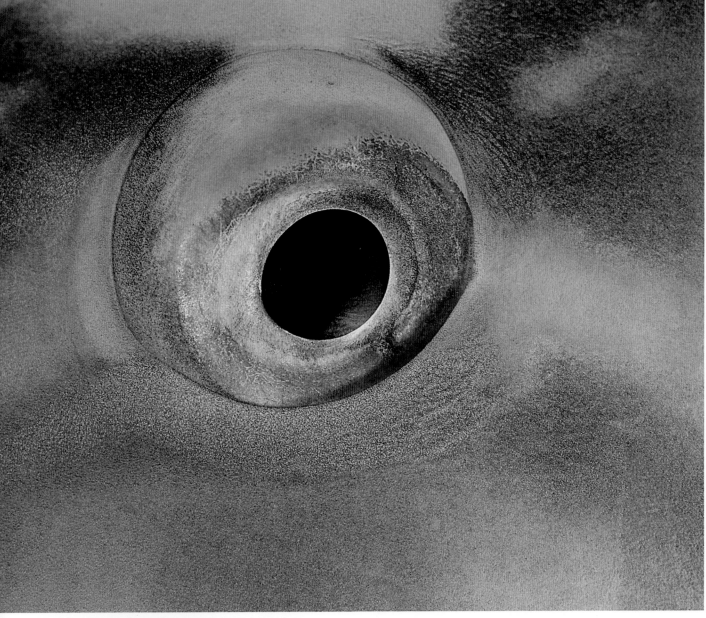

*82-83 Detail of a parrotfish's eye (Scarus sp.).*

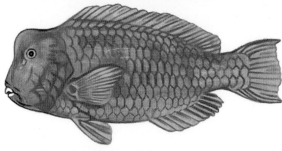

**Bumphead parrotfish**
*Bolbometopon muricatum*

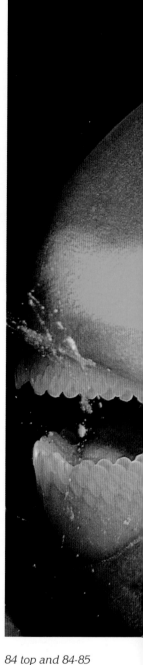

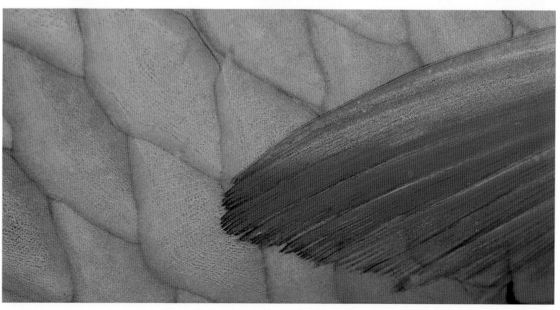

*84 top and 84-85
Rusty parrotfish
(Scarus
ferrugineus).*

*84 center
Steephead
parrotfish
(Scarus gibbus).*

*84 bottom
Detail of
parrotfish's scales.*

As any scuba diver soon discovers, the underwater world is anything but silent. Creaking noises, whistles, and thumping roars are common sounds under the surface, but the most easily recognizable of them all are the sharp cracks produced by the *Scaridae* or parrotfish as they tear away at the coral with their powerful beaks. Exceedingly colorful and perhaps far more changeable than the *Labridae*, since their colorings can change with age, gender, and season of the year, the Scaridae have powerful bodies, flattened sidewise, covered with large scales. More than by these characteristics, however, the parrotfish (genera *Scarus, Bolbometopon*, etc.) can be easily recognized by their mouths, featuring large and prominent teeth that have been transformed into dental plates in the form of a beak, perfectly suited to

breaking away pieces of coral; the parrotfish in fact feed on polyps. At regular intervals, then, it is possible to observe around the parrotfish the sudden blossoming of a little white cloud, which quickly dissolves. The cloud consists of the undigested remains of the coral, ground up and transformed into a very fine coral sand. Like the *Labridae*, the *Scaridae* too are diurnal fish.

At sunset, each specimen takes refuge in its den or hideaway, and begins to secrete a transparent mucous covering, a sort of cocoon that entirely envelopes the fish, and dissolves the following day. The function of this covering (which not all *Scaridae* possess) is believed to be that of preventing nocturnal predators from discovering the sleeping parrotfish by their sense of smell.

# SURGEONFISH

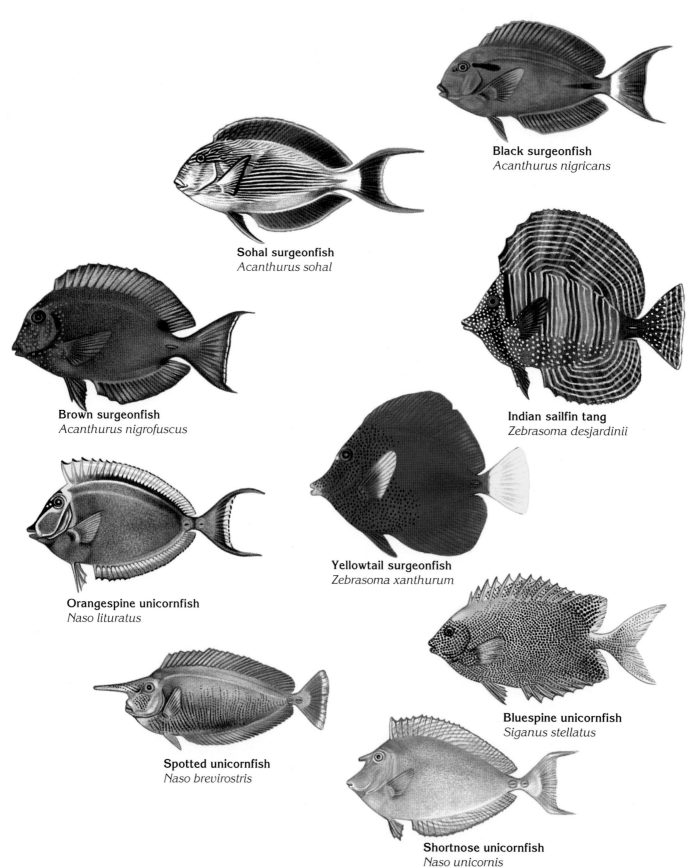

**Black surgeonfish**
*Acanthurus nigricans*

**Sohal surgeonfish**
*Acanthurus sohal*

**Brown surgeonfish**
*Acanthurus nigrofuscus*

**Indian sailfin tang**
*Zebrasoma desjardinii*

**Orangespine unicornfish**
*Naso lituratus*

**Yellowtail surgeonfish**
*Zebrasoma xanthurum*

**Bluespine unicornfish**
*Siganus stellatus*

**Spotted unicornfish**
*Naso brevirostris*

**Shortnose unicornfish**
*Naso unicornis*

Oval in shape, with broad dorsal and anal fins, and with tails that are often sickle-shaped, with elongated lobes, the *Acanthuridae*, or surgeonfish are quite common in the central reef area, where the lighting is brightest and where the algae eaten by these fishes grow most abundantly. One distinctive characteristic shared by all surgeonfish, and which has given the group its strange name, is the presence of remarkably sharp spines, as sharp as a scalpel, set along the sides of the caudal peduncle. Observing the peduncle with particular care one can make out an area of contrasting color with sharp spines generally movable, which can be raised in self defense all pointing forward. In general, they are used to settle territorial disputes, but more as a method of intimidation and threat than as an actual weapon. These spines also pose a potential danger for humans, and it is best to keep this in mind in the presence of large schools of *Acanthuridae*.

*Sohal surgeonfish*
(Acanthurus sohal).

*87 bottom*
*Bluespine unicornfish*
(Naso unicornis).

*87 center*
*Blacktongue*
*unicornfish*
(Naso hexacanthus).

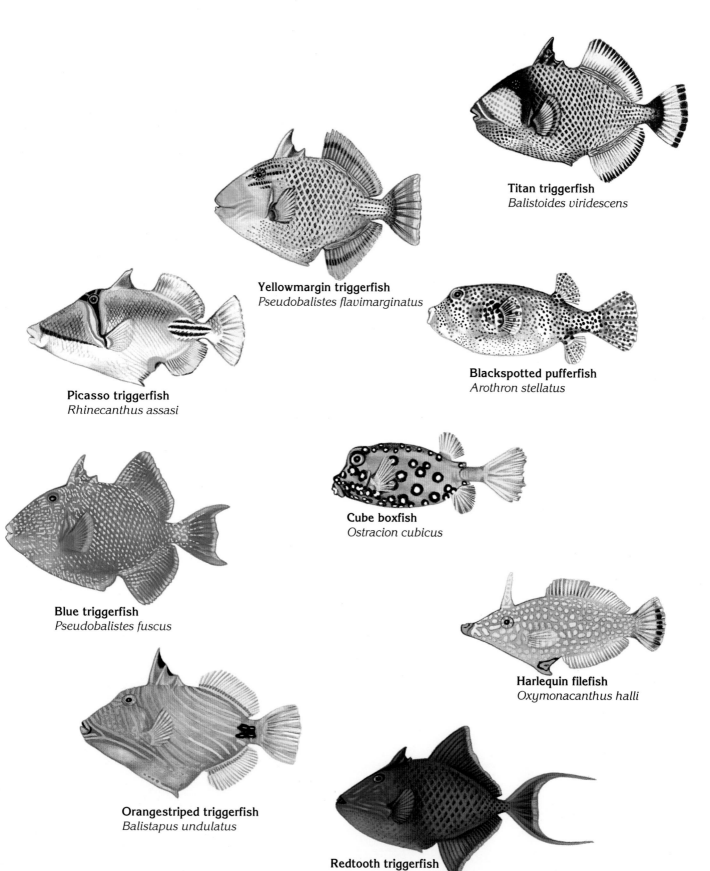

**Titan triggerfish**
*Balistoides viridescens*

**Yellowmargin triggerfish**
*Pseudobalistes flavimarginatus*

**Picasso triggerfish**
*Rhinecanthus assasi*

**Blackspotted pufferfish**
*Arothron stellatus*

**Cube boxfish**
*Ostracion cubicus*

**Blue triggerfish**
*Pseudobalistes fuscus*

**Harlequin filefish**
*Oxymonacanthus halli*

**Orangestriped triggerfish**
*Balistapus undulatus*

**Redtooth triggerfish**
*Odonus niger*

*88-89*
*Yellowspotted*
(Chilomycterus
spilostylus).

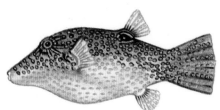

**Pearl toby pufferfish**
*Canthigaster margaritata*

**Burrfish**
*Diodon hystrix*

Large, slightly flattened on the sides, and covered with small but extremely strong bony plates, the triggerfish (*Balistidae*) are distinguished by two dorsal fins. The first of these fins gave the fish their name, as it is made up of three rays that can be intertwined with a triggering mechanism that locks the longest ray into place. This mechanism goes into operation when a triggerfish takes shelter in a grotto or fissure, locking it into place so that it becomes impossible to extract it from its lair. Another notable characteristic of the triggerfish consists of its dentition, equipped with powerful chisel-teeth, fastened solidly to jaws moved by particularly powerful muscles, with which these fish can easily crush the shells of the crustaceans, molluscs, and sea urchins on which they feed. They use a remarkable technique to catch sea urchins. Fearful of the sharp spines with which the sea urchins defend themselves, the triggerfish may seize two of the spines, and with them drag the sea urchin upward, release it, and then attack it from its unprotected lower side as it drops slowly through the water; they also direct powerful jets of water at the base of the sea urchin to overturn it for the same purpose. The triggerfish, lastly, are very territorial and it is advisable to keep a safe distance to avoid being attacked, especially when they are guarding their nests.

90-91
*Orangestripped
triggerfish*
(Balistapus
undulatus).

91 top
*Titan triggerfish*
(Balistoides
viridescens).

91 bottom
*Red tooth
triggerfish*
(Odonus niger).

92 top
Masked pufferfish
(Arothron
diadematus) *and*
*blacktip grouper*
(Epinephelus
fasciatus).

92 center and 80-81
Blackspotted puffer
(Arothron stellatus).

92 bottom
Cube trunkfish
(Ostracion cubicus).

Powerful teeth are also distinctive characteristics of globefish, porcupinefish, and boxfish (*Ostraciidae*). The latter have a rigid body, made up of bony plates that are often hexagonal, from which tiny fins project, their swirling motion moving these rigid fish through the corals like tiny helicopters. The globefish (*Tetraodontidae*) and porcupinefish (*Diodontidae*), on the other hand, share the ability to inflate themselves when threatened. This feature, which always interests and attracts divers, is exhausting for the fish; therefore, one should not touch them or attempt to provoke them into inflating themselves, as this will exhaust and cause them harm.

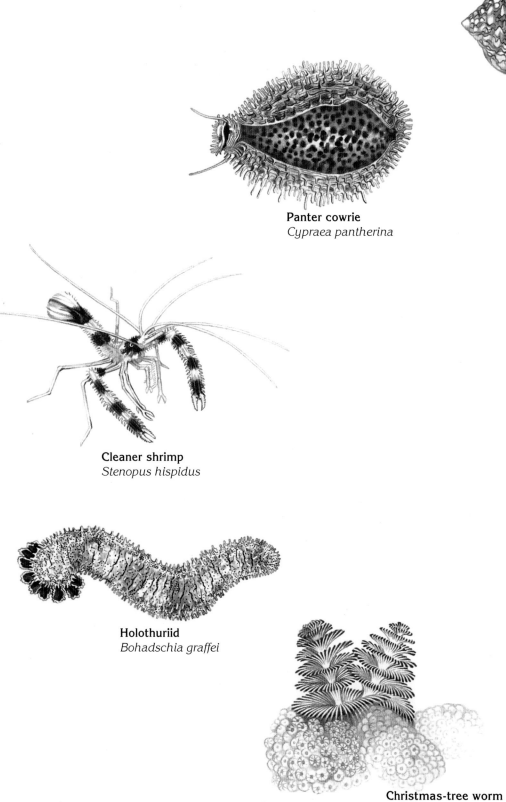

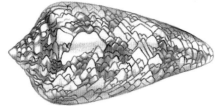

**Cone shell**
*Conus textile*

**Panter cowrie**
*Cypraea pantherina*

**Cleaner shrimp**
*Stenopus hispidus*

**Holothuriid**
*Bohadschia graffei*

**Christmas-tree worm**
*Spirobranchus giganteus*

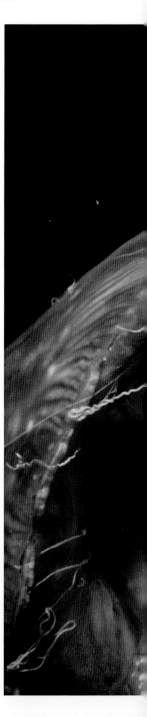

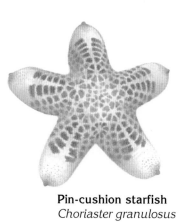

**Pin-cushion starfish**
*Choriaster granulosus*

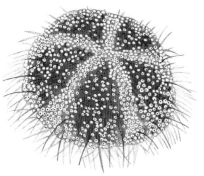

**Pin-cushion urchin**
*Asthenosoma varium*

**Starfish**
*Fromia sp.*

95 *Ermit crab*
(Dardanus sp.).

*96 Spanish
dancer*
(Hexabranchus
sanguineus).

*97 left
Christmas tree
worm*
(Spirabranchus
giganteus).

*97 top right Crown
of thorns starfish*
(Acanthaster planci).

*97 center right
Giant clams*
(Tridacna
maxima).

*97 bottom right
Chromodoris*
(Chromodoris
quadricolor).

Extremely common, as well, are the mollusks and echinoderms. Among the mollusks, there are at least three species that can be considered emblematic of these sea beds: the Spanish dancer (*Hexabranchus sanguineus*), the triton (*Charonia tritonis*), and the clam (*Tridacna maxima*). The first-named species is the largest and most spectacular nudibranch to be found in the Red Sea; in some areas it can attain a length of as much as 40 centimeters. Reddish in color, with white edges and large feathery gills, this nudibranch - with the delicacy of its movements - can capture the imagination of anyone having the good fortune to see and watch it swim through the water. Equally large is the triton, the chief predator of the crown of thorns starfish (*Acanthaster planci*), which can destroy in a very brief time as much as a square meter of coral reef.

Clams are too well known to be described, though we need to point out that their reputation as traps for scuba divers is quite exaggerated. Certainly, if they are touched, they do react by closing their valves, but the movement involved is so slow that anyone would have time to avoid being caught.

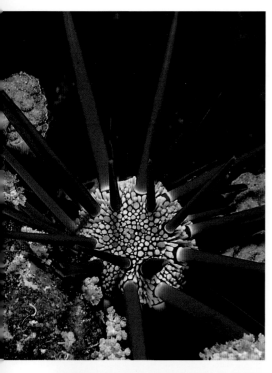

Among the Echinoderms, the sea urchin is the best known but also the most dangerous. Alongside the innocuous pencil urchin, which by day remain hidden in the fissures, and the sand dollars, buried in the bottom silt, there are at least two other species that is worthwhile to learn to recognize: the diadema urchin (*Diadema setosum*) and the pincushion urchin (*Asthenosoma varium*).

The former has exceedingly long spines, as thin as glass needles, which break at the slightest impact. Coated with poison, they can cause painful wounds. Oddly enough, this sea urchin is endowed with sensitive cells that are similar to eyes. It is enough to cast a shadow on one of these creatures to see it react, orienting all its spines immediately in one's direction. Apparently more harmless, with its reddish color upon which rounded white formations stand out, is the pincushion urchin.

Actually, however, the white formations hide a dangerous feature, as they are packed with poison and connected to sharp spines.

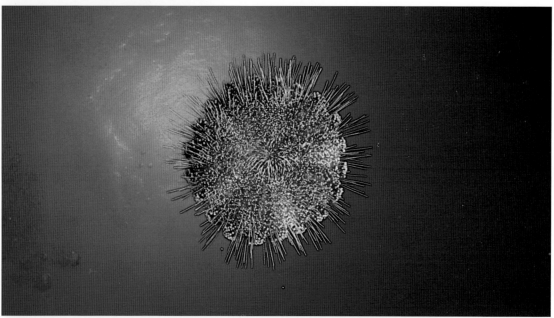

*98 top*
*Slate pencil urchin*
(Heterocentratus mammilatus).

*98 center and bottom*
*Pincushion urchin*
(Asthenosoma varium).

*99 Starfish*
(Fromia sp.).

Quite common, especially where constant currents prevail, are the other Echinoderms, such as the Crinoids or the sea lilies, with their numerous feathery arms spreading out from a small body that remains fastened to corals or to gorgonians through articulated, prehensile tentacles. Lastly, we cannot overlook the crustaceans. Lobsters, crayfish, crabs, and hermit crabs move to and fro across the sea bed, especially by night, working as efficient and thorough scavengers. Some of them, on the other hand, establish relationships of close-linked symbiosis with larger animals (sea anemones and fish), cleaning them of parasites and organic residue in exchange for protection and food. In general, these species are smaller, and it takes some practice to see them, but real diving means paying attention even to the smallest organisms.

*100 top*
*Feather stars*
*(Heterometra sp.).*

*100 bottom left*
*Mimetic*
*spider crab*
*(Majidae fam.)*

*100 bottom right*
*Humpbacked*
*shrimps*
*(Hyppoliysmata*
*grabham).*

*101 Cleaner shrimp*
*(Periclimenes sp.).*

# THE REEF BUILDERS

**Mountain coral**
*Porites sp.*

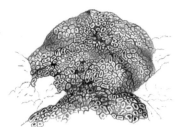

**Rose coral**
*Lobophyllya sp.*

**Mushroom coral**
*Fungia sp.*

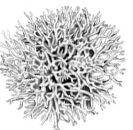

**Spiny row coral**
*Seriatopora hystrix*

**Brain coral**
*Platygyra sp.*

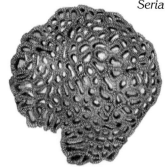

**Honeycomb coral**
*Favites sp.*

**Soft coral**
*Dendronephthya sp.*

**Red cave coral**
*Tubastrea sp.*

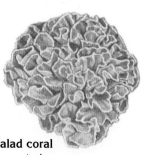

**Serpentine salad coral**
*Turbinaria mesenterina*

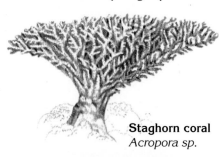

**Staghorn coral**
*Acropora sp.*

**Leafy coral**
*Pachyseris sp.*

**Sea fan**
*Subergorgia hicksoni*

*102-103*
*Jewel fairy basslet*
(Pseudanthias
squamipinnis)
*swim amidst soft*
*corals*
(Dendronephthya
sp.).

**Fire coral**
*Millepora dichotoma*

**Raspberry coral**
*Pocillopora sp.*

Even more complex and varied than the world of the fish is the universe of the invertebrates, which dominates in terms of numbers the floor of the Red Sea. The coral reefs built by the incessant and slow activity of dozens of different species of corals are the very foundation of underwater life. Encouraged by crystal-clear waters, warm and rich in nutrition, the tiny polyps that make up the colonies of the *Acropora* or the Madreporarians multiply, synthesize calcium carbonate from sea water and erect fantastic constructions, in an apparent chaos that actually meets exceedingly precise ecological requirements.

*104-105 Detail of the tips of branches of coral. (Acropora sp.).*

*105 top Fire corals (Millepora dichotoma).*

*105 bottom Distinctive picture of the Red Sea with hard and soft coral formations (Sarcophyton sp.).*

For one species, light may be the limiting factor; for other species it may be the degree of exposure to the wave action or to the currents that limits growth. The apparent similarity to petrified trees and bushes, or in some cases to large rocks, should not fool the diver. Though the polyps may be tiny and almost invisible, or else visible only by night, they are the fundamental component of the coral, from which they extend their tentacles into the water in a continual search for food. Easier to observe are the very colorful soft corals, similar in appearance to little translucent trees, which expand by night in luxuriant forms to intercept with greater efficiency the clouds of plankton that sweep by on the currents. Without a rigid skeleton, and therefore described as soft, the Alcyonarians are most common in the upper 15 meters from the surface, where there are many species — more or less encrusting — that at first glance look like large sea anemones with short but numerous tentacles, or formations of moss, in some cases covering many square meters of sea floor. Also very common are the gorgonians, generally quite similar to those found on the floor of the Mediterranean, but in some cases quite different, and capable of taking on a threadlike shape, straight, with a slight spiralling bend, or with a terminal curl, that distinguishes this creature from the so-called whip corals.

*106 top left*
*A group of tubular sponges*
(Siphonochalina sp.).

*106 center left*
*A series of encrusting*
Anthozoa *with expanded tentacles.*

*106 center right*
*An expanded polyp of* Tubastrea.

*106 bottom left*
*A colony of gorgonians*
(Plexauridae fam.) *offers support to a number of crinoids.*

*107 A giant gorgonian sea fan*
(Subergorgia hicksoni) *.*

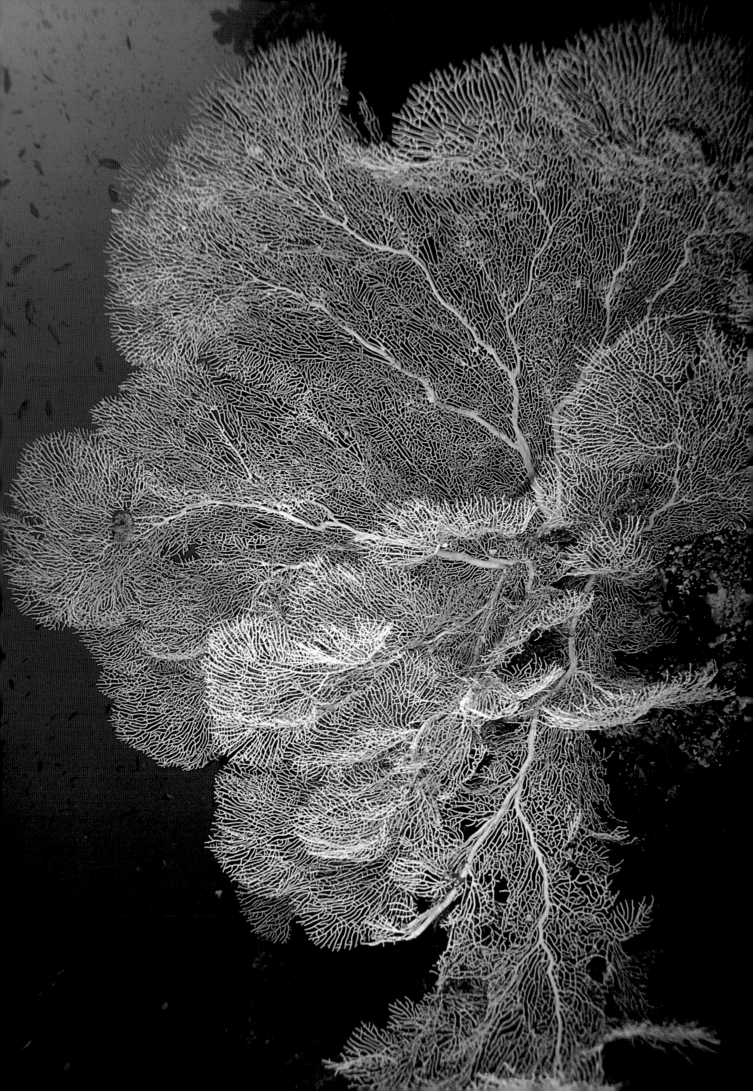

# The Maldives

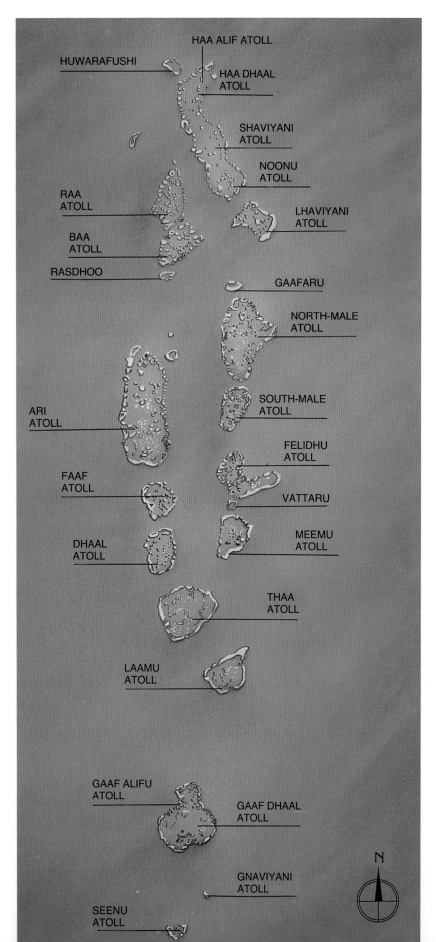

HUWARAFUSHI

HAA ALIF ATOLL

HAA DHAAL ATOLL

SHAVIYANI ATOLL

NOONU ATOLL

RAA ATOLL

LHAVIYANI ATOLL

BAA ATOLL

RASDHOO

GAAFARU

NORTH-MALE ATOLL

ARI ATOLL

SOUTH-MALE ATOLL

FELIDHU ATOLL

FAAF ATOLL

VATTARU

DHAAL ATOLL

MEEMU ATOLL

THAA ATOLL

LAAMU ATOLL

GAAF ALIFU ATOLL

GAAF DHAAL ATOLL

GNAVIYANI ATOLL

SEENU ATOLL

N

**Texts**
Angelo Mojetta

**Translations**
A.B.A. s.r.l., Milan

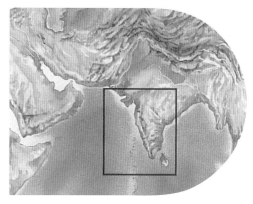

# Contents

INTRODUCTION                     page  110

- THE LAGOONS                    page  122
- THE THILAS                     page  130
- THE PASSES                     page  136
- REEF FISH                      page  144
  - Sand dwellers                page  144
  - Grotto dwellers              page  150
  - Coral dwellers               page  158
  - Coral predators              page  166
  - Lurking in the coral         page  174
  - The groupers' territory      page  182
  - In the coral jungle          page  188
  - Horns, spines and teeth      page  194

*109  Razor fish
(Aelodiscus strigatus).*

# INTRODUCTION

## THE INDIAN OCEAN AND THE ORIGINS OF THE MALDIVES

In view of their unique characteristics the Maldives, which have been described with countless adjectives and compared to emeralds studding the sea or liquid transparencies in every shade of blue and green, have never ceased to fascinate travellers, explorers and divers. This extraordinary archipelago is surrounded by the Indian Ocean which, although it is the smallest ocean on our planet, features various kinds of reef.

*110 These aerial photos illustrate three characteristic coral formations: an atoll (top), some arc-shaped elongated reefs (center) and a ring-shaped faru (bottom).*

There are atolls, barrier reefs and fringing reefs, reefs which emerge above the surface of the sea and others which are wholly submerged. The locations and quantities of hard coral formations in this ocean are similarly varied. There are few or no reefs along the coasts of Somalia, the Indian sub-continent and Malaysia, for example, whereas they are highly developed in the Red Sea, the Persian Gulf, Madagascar, Sumatra, the Seychelles, etc. Barrier reefs are quite rare, whereas fringing reefs are more common, at least along the continental coasts of Africa and Asia and near the archipelagos of the Seychelles and the Nicobar, Comoro and Mascarene Islands. Atolls are more developed and characteristic of the central part of the Indian Ocean. The distribution and size of the reefs listed above depends largely on the geological history of the ocean in which they are formed and on the circulation of the ocean currents, which are strongly influenced by the monsoons. Most of the seabeds in the Indian Ocean lie at depths ranging between 4,000 and 5,000 meters, and support impressive rock formations similar to mountain chains which stretch for thousands of kilometers, mainly running north to south.

The geological history of the Maldives can be said to have begun some 200 million years ago, when the gigantic block called Pangea, which then joined all the land above sea level, gradually began to break up. India, Australia, Madagascar and the Antarctic began to drift away following the gradual opening up of the Atlantic Ocean and the emission of lava, which took place between southern Africa and Antarctica and then between Antarctica and Tasmania.

Just over 150 million years ago, the Indian Ocean also began to form. This process, which was initially very slow, gradually accelerated due to the eruption of large amounts of basalt lava between north-east India, Antarctica and Australia.

The Indian Plate began to move north-east, breaking off from Africa and Madagascar. At about the same time, when India was separated from the Seychelles and the Arabian Sea opened up, another great episode of basalt eruption took place (so intense that some geologists believe it caused the extinction of the dinosaurs) which produced the Deccan plateau, a layer of basalt over 2,000 meters thick, covering an area of almost a million square kilometers.

These events were associated with the passage of India over one of the "hot spots" of the earth, fixed areas of its crust below the lithosphere which feature intense volcanic activity and high production of magma because the abundance of water lowers the melting point of the rock. The location of this hot spot, which played an important part in the history of the Maldives, has been identified near the volcanic island of Réunion. Geological surveys have revealed that the area now occupied by the Chagos, Maldive and Laccadive archipelagos passed this way. Their passage at this point caused the formation of a series of volcanoes which followed India as it drifted, marking its route north until it collided with Asia some 50 million years ago, throwing up the chain of the Himalayas and defining the shape of the Indian Ocean.

The chain of volcanoes which formed the ancient Maldives, believed to be similar to the present Hawaii, slowly sank below the surface of the sea, while coral formations became thicker and grew at a fast enough rate to make up for this sinking. Some experts believe that in addition to these phenomena there have been major variations in sea level in the past 40,000 years, but although this hypothesis is

acceptable for other coral areas, it does not seem applicable to the Maldives, where the ocean appears to have maintained constant levels in the past.

The growth of the Maldives above a chain of volcanoes was demonstrated once and for all by test holes drilled during an oil prospecting campaign.

A deep well dug in the North Male Atoll reached a volcanic basement at a depth of 2,100 meters, colonised by coral

dating back 53-38 million years. Thus the long northern atoll of Thiladhunmathi-Miladhummadulu (155 kilometers long by 30 kilometers wide) may have formed at the peaks of a number of volcanoes, while the giant atoll of Huvadhoo and the smaller Addu Atoll in the south of the archipelago may indicate the presence of two large, deeply submerged volcanoes.

*111 The coral sand surrounding the atolls has an unmistakeable pale color, with the result that the waters bordering the island always have a characteristic turquoise shade.*

# THE BIRTH OF THE ATOLLS

The Maldive archipelago, which stretches for nearly 900 kilometers, consists of a twin series of atolls standing on a wide submerged platform located at a depth of 270-500 meters, bounded by waters which reach a depth of 2,500 meters on the east side and approximatively 4,000 meters on the west side. An atoll (this name is derived from the Maldivian *atholu*, which

The five drawings illustrate the phases leading to the formation of an atoll. It originates from a series of fringing coral reefs standing on rocky shoals surrounding a volcanic island emerging from the sea (A). Gradual lowering of the seabed following subsidence and erosion of the volcano (B) increases the size of the fringing reefs, which come to form a more or less continuous barrier (C). The incessant growth of coral enables the reef to continue its development, and compensates for the lowering of the volcanic rock (D) until, when the island is totally submerged, a lagoon is formed, surrounded by a circular reefs containing channels which allow water to be exchanged between lagoon and sea (E).

*112 These pictures show two different stages in the development of an atoll.*

actually indicates both natural atoll and administrative unit) is defined as a coral formation surrounding a circular central lagoon, whose size can range from a few kilometers to dozens. For example, the Huvadhoo Atoll in the south Maldives is around 68 kilometers long and 55 kilometers wide. These hard coral constructions are usually found in very deep ocean waters, on the sites of ancient submerged volcanic islands. Because of their characteristics, atolls, more than any other kind of coral reef, have attracted the attention of researchers, who have mainly endeavoured to explain their origin. One of the first to study atoll formation was Charles Darwin, whose theory is still the most generally accepted, and is now supported by detailed geological and palaeontological surveys. Atolls originate from a series of fringing reefs standing in rocky shoals surrounding a volcanic island emerging from the sea. The gradual lowering of the seabed following subsidence seems to allow the coral formations to increase in thickness and develop a more or less continuous reef which, when the island is totally submerged, eventually surrounds a lagoon.

The lagoons in the Maldives have depths ranging from 20-30 meters to the 84 meters of

The drawing above illustrates an imaginary longitudinal section through an island of the Maldives and its reefs. The section shows an atoll situated between the ocean, 2,000-3,000 meters deep, and the inner lagoon of the atoll, which can be up to 80 meters deep. In addition to the atoll lagoon, there are other smaller lagoons around the island protected on both sides by a coral reef, which is richer and more varied in the outer lagoon areas. The dominant feature is the island, generally composed of hard calcareous rock, on which lush vegetation consisting of coconut palms, pandanus and other tropical plants grows.

the Huvadhoo Atoll. However, the structure of the Maldivian atolls is more complex than it might seem from this simple description, which explains the general form of these coral formations but not their components. One of the unusual features of the Maldives is the composition of its atolls, which are formed by a number of small atolls called *faru* or *farus*, separated by channels or passes *(kandu)* of various widths, which enable water, fish, nutrients and waste matter to be exchanged between the open sea and the lagoons.

*113 This photo clearly shows a pass, the channel which allows water to flow into the atoll so that its waters are continually changed.*

The main difference between an atoll and a *faru* relates not so much to their size, as atolls smaller than a *faru* exist, but to the type of seabed on which they stand. Atolls stand on ocean beds, whereas a *faru* is a coral formation which rises from the bed of the atoll. A *faru* consists of a raised part, which may give rise to a verdant island, and a shallow lagoon (*falhu* or *vilu*) surrounded by coral formations which face the ocean on one side and the lagoon of the atoll on the other. As will be seen from looking at a map of the Maldives, the morphology of the atolls changes considerably as we proceed further south, and the same applies to the shape of the *farus*, the depth of the lagoons and the percentage occupied by reefs emerging above the surface of the sea. These differences, which lead to the formation of ribbon and horseshoe reefs, and reefs composed of elongated or ring-shaped *farus*, are generally explained by variations in the environmental conditions which influence the growth of the coral; in view of the fact that the Maldives stretch for 900 kilometers, this theory seems reasonable.
In addition, the action of the monsoons seems to have a major effect on the shape of the *farus* and the existence of a double chain of islands. Monsoons are periodic winds which blow from south-west in summer and from north-west in winter, affecting the coasts of the Maldives alternately and reversing the dominant currents. The general result of the seasonal alternation of the currents and the prevalent wave direction is symmetrical development of atolls, because the changing and alternating ocean conditions cause the coral to grow in all directions, not in one prevalent direction as is the case with other reefs on which constant winds blow in only one direction.

*114 top The vast blue reaches of the Indian Ocean are studded with separate reefs. Their shape and development always depend on the action of the currents and the motion of the waves. There are even seasonal alterations of the shape of the atolls and one of these is due to the movement of beach sand from one location of the island to another. Another seasonal manifestation is the appearence or disappearence of sandbanks in some parts of the Maldives.*

*114 bottom This aerial photo shows the linear development of ribbon reefs, which form a kind of continuous barrier separating the deep waters of the ocean from the shallow waters of the lagoon.*

A

The action of the waves, currents and winds considerably influences the shape of the Maldivian atolls and the reefs surrounding them; their appearance gradually changes from north to south, where large circular atolls tend to predominate.

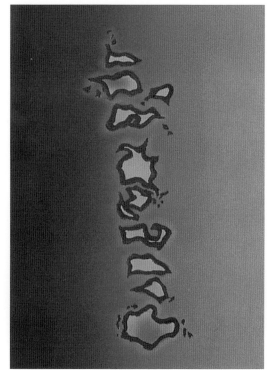

D

On the basis of shape, the following types of reef can be identified:
A) reefs separate from one another, which tend to be arc-shaped;
B) ring-shaped reefs with large inner lagoons, which are clearly separated from one another;
C) ribbon reefs, which form an almost continuous barrier between ocean and lagoon;
D) reefs which are quite separate but parallel to one another, and perpendicular to the imaginary boundary between ocean and lagoon.

A

D

B

C

B

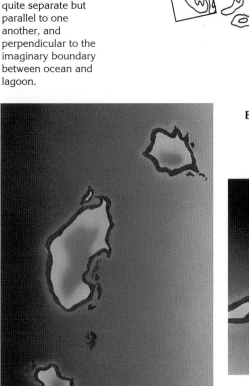

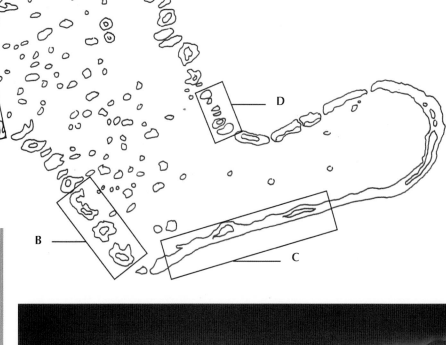

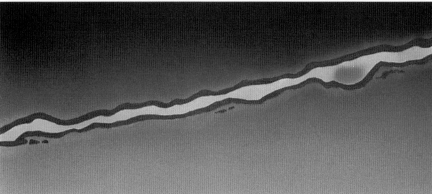

C

## THE CORAL BEDS OF THE MALDIVES

The seabeds of the Maldives feature the richest coral formations in the Indian Ocean. Research has enabled over 66 genera and over 100 different species of hard coral to be identified. This great variety is due to the favourable environmental conditions in the region, but it would be incorrect to assume that coral grows everywhere and in a uniform manner. The different environments that can be identified in an atoll usually feature associations of hard coral, only some of which

are really predominant. Taking one of the green islands dominated by palm and pandanus trees as our starting point, we can trace two itineraries based on the most easily recognisable environments, one leading to the atoll lagoon, and the other to the ocean.

The classic sub-division of the first itinerary is constituted by a sequence of four separate zones, whose borders and size can vary widely from one part of the Maldives to another: an inner zone, a mixed-coral zone, a zone

*116 top Numerous dead corals are found in the lagoon.*

*116 bottom This butterflyfish (Heniochus dephreutes) can be found in the passes, where there is the greatest flow of waters rich in nourishment.*

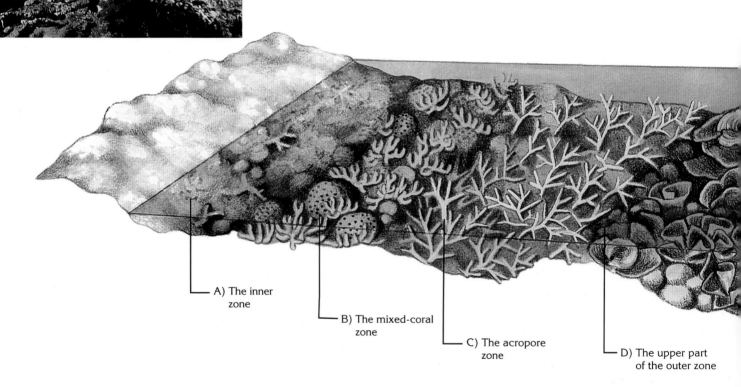

A) The inner zone

B) The mixed-coral zone

C) The acropore zone

D) The upper part of the outer zone

In this drawing it is clearly shown the area between the island beach and the reef facing the atoll lagoon. This area is divided into a series of belts, in which the species best suited to the local environmental conditions predominate.

A) The inner zone, which starts at the water's edge, is dominated by sand, and dead coral is also present.
B) The mixed-coral zone mainly features acropores (e.g. *Acropora digitifera*), pocillopores, blue coral and massive coral.

C) The acropore zone is dominated by *Acropora formosa*, a species of acropore with numerous antler-shaped branches.
D) The upper part of the outer zone is heralded by a belt of mixed brain or foliaceous coral.
E) The slope of the

reef is the richest part, with tabulate acropore, laminar madrepore (*Echinopora sp.*), fungus coral (*Fungia sp.*, *Herpolitha sp.*), massive honeycomb or brain-shaped colonies and clinging corals.

*117 top left*
*A flourishing colony of delicate soft corals has developed on a stretch of hard corals.*

*117 top right*
*An impressive umbrella acropore has grown perpendicular to the reef, perhaps to exploit the plankton-rich currents.*

*117 bottom*
*A coral grouper (Cephalopholis miniata) stands out among the branches of an unusual coral composition.*

E) The slope
of the reef

dominated by coral of the species *Acropora formosa*, and an outer zone.

The inner zone, whose size varies, depending on the shape and size of the *farus*, begins at the water's edge and is dominated by sand, with an almost total absence of corals. The first part is marked by undulations in the sand of the seabed caused by wave movements which cloud the water. It is followed by a zone containing pieces of dead coral or massive blocks of hard coral firmly cemented to the seabed, containing crevices worn away by the waves in which molluscs and small echinoderms are found. In some islands, this zone may also be covered with the long, ribbon-shaped leaves of marine phanerogams of the *Thalassia* genus.

In the mixed-coral zone, the coral gradually increases as we move further from the shore. The first corals to appear are acropores (e.g. *Acropora digitifera*, recognisable by its short finger-like branches) which mainly colonise old dead coral and can grow in surface water and withstand brief emergences at low tide. Mingling with these hard corals are the pocillopores, some colonies of the so-called "blue coral" *(Heliopora sp.)* and small formations of massive coral *(Goniastrea sp.)* with regularly-shaped corallites distinguished by high walls, and rare specimens of fungus hard coral *(Fungia sp.)*.

The next zone is dominated by *Acropora formosa*, a species of branched acropore with numerous staghorn-shaped branches which are sometimes very long and sharp. This hard coral is particularly common in deeper waters, and it is by no means unusual to find it in the channels between one sandy shoal and the next. Depending on the prevalent conditions and lesser or greater hydrodynamics, this species is predominant or replaced by massive corals like the *Goniastrea* and brain hard corals.

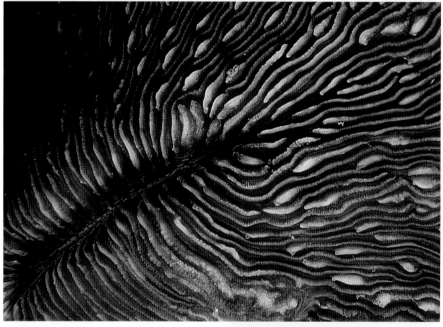

The closeness of deeper waters and conditions more favourable to the development of coral coincides with the outer zone, heralded by a belt of mixed coral in which the predominant species are massive, hemispherical forms sometimes over a meter long, weakly branched or foliaceous.

The brain hard corals and the slender but wide, flattened formations of the echinopores are easily recognized. The coral belt acts as a prelude to the actual reef, which slopes down towards the deeper part of the lagoon. The upper edge of the reef, which is directly exposed to the waves, is one of the richest areas, only 10% of which is devoid of coral. The most prevalent types are the leaf-shaped or tabulate species such as *Acropora hyacinthus*, which forms wide, horizontal plates consisting of short, finger-like branches growing very close together. These species are sometimes also recognisable by the pink or bluish color of the growing edges. *Echinopora laminosa* is frequently found; as its name suggests, it forms laminar colonies with protruding corallites and undulating edges which in some cases are bent like pipes. This type of coral continues to an average of 10-15 meters, depending on the slope of the reef and the play of currents and waves.

At these depths the hard coral formation begins to be interrupted by sandy zones abounding in the fungal and discoid corals of the *Fungia* genus and the more oval, elongated corals of the *Herpolitha* genus.

The community of deep corals includes encrusting species belonging to the *Pavona* and *Pachyseris* genera, massive hard corals (*Favites, Favia, Goniastrea* and *Porites* genera) and branched hard corals, the *Acropora* genus once again being predominant. The lower limit of the reef is marked by the continuous flat, sandy area

*118 top Red stinging coral is particularly fragile and delicate, yet it flourishes at great depths.*

*118 center Macrophotography enables the laminar structure of a* Fungia *coral to be observed.*

*118 bottom The branches of black corals (Anthipatharia), similar to densely-packed fronds, house numerous life forms which seek protection and nourishment among them.*

of the lagoon bed, which can be as much as 60-80 meters deep. The area which stretches from the island to the sea also presents a type of zoning which enables at least four different environments to be identified: a platform, a debris area, a series of channel-like depressions and a formation dominated by algae. The initial platform, which at some points is uncovered at low tide, consists of rock of coralline origin covered with a thin layer of sand, and sometimes marine plants. Coral is rare, and represented by a few colonies of blue coral *(Heliopora sp.)* and larger formations of *Porites.*

The platform is bounded on the seaward side by a belt of coral debris of varying size (up to 1 meter in diameter), forming long strands in the direction of the platform, thought to be produced by the action of the tides. Here, the seabed is more irregular; flat areas alternate with pools in which encrusting corals or corals with short, thick branches may grow (*Porites, Acropora* and *Pocillopora* genera).

The next zone, featuring a series of wide, shallow channels which are covered with water even at low tide, is very unusual. The seabed consists of large pieces of hard coral debris, encrusting algae and some corals, including stinging corals (*Millepora sp.*), which create an environment full of crevices where molluscs and echinoderms hide away.

The outer edge preceding the actual reef is dominated by calcareous encrusting algae which create a belt on which the waves break; they are only rarely out of the water. The algal reef is intersected with channels produced by the waves; water pours in rapidly through these channels, then flows out gradually in the opposite direction in dozens of rivulets which run through the labyrinth created by the algae. There are also small colonies of hard coral, stinging coral (most frequent at the edges of the channels) and *Zoantharia.*

The outer slope of the reef on the seaward side consists of a more superficial part dashed by the waves, and a submerged sloping terrace followed by an embankment which slopes steeply down to depths of the ocean. The area nearest the surface consists of projecting rocks intersected by deep clefts hollowed out by the masses of water moved by the motion of the waves, which are channelled violently here, limiting the development of the coral to a few massive branches of palmate acropore, and continually stirring up the sediments on the sea bed. On some particularly exposed reefs, the movement of the ocean waves is felt even at a depth of 20-25 meters. Below the belt on which projecting rock alternates with hollows is a sloping terrace with coral formations dominated by palmate and ramified acropore, mixed with colonies of *Echinopora*, whose structures, though leaf-shaped, are thickened to withstand the action of the sea. Areas containing a wealth of sand and dead coral debris open up among the corals. Colonies of alcyonarians, with gorgonians and tree-shaped colonies of *Tubastreae*, which are greenish-black by day and golden yellow by night, when the polyps are fully expanded, start to become frequent at a depth of about 20 meters. Coral growth is extremely slow and even the slightest touch by a snorkeller or a diver's hand or fins could rob the reef of many years of growth. They can help to ensure the survival of the reefs with the care and respect they deserve. This would contribute to preservation of the underwater paradise and heritage of the Maldives for future generations.

*119 The alcyonarians have different characteristic shapes: this illustration shows the typical fans of the gorgonian sea fans at the top, the elongated branches of whip corals at the center, and the leafy white branches of soft corals at the bottom.*

# Under the Waters of the Maldives

Stretching across the Indian Ocean for nearly 900 kilometers, with a proportion between dry land (under 300 square kilometers) and marine environment which can be described as infinitesimal, the Maldives are

*120 top  A pair of black-footed clownfish (Amphiprion nigripes) lives in constant contact with the sea anemone which, with its stinging tentacles, guarantees continuous protection.*

*120 center Feathery yellow crinoids have colonised a block of coral.*

*120 bottom The valves of this Tridacna (Tridacna maxima) are partly closed in the grip of the madrepore.*

a paradise of aquatic life, containing over a third of the coral fish found in the entire Indian Ocean.

The main reason for this abundance lies in the variety of environments in the archipelago, where as well as barrier reefs rising from a depth of 2,000-3,000 meters up to the surface, dashed by waves and fast currents, there are also quiet lagoons fringed by white beaches. Always changeable, uneven and intersected by crevices which turn into tunnels and canyons, reefs provide the ideal habitats, which meet the requirements of every species and the communities they form. Some find them the ideal refuge for hiding from predators, while others lie in wait there, or hide their eggs or young there.

In some cases, evolution has led to the establishment of extraordinary interspecific relationships, not only between animals belonging to the same zoological group, as in the case of the cleaner fish and their customers, but also between organisms which are

phylogenetically very different, such as clownfish and anemones, or gobies and crustaceans.

The complex relationships between fish and environment can only be appreciated after repeated dives. The chaotic comings and goings of the anthias, for example, are caused by the movement of masses of water and the masses of plankton associated with them. These fish are also influenced by the passage of larger animals as well as divers and by the territorial relationships established between the dominant territorial males and their harems of females. It is not only their behaviour which astonishes divers. On every dive, and even merely during a brief snorkelling excursion, you can't help but notice the different shapes of the fish: serpentine (moray eels, pipefish and trumpetfish), tapering (groupers, wrasses, tuna fish and jacks), spherical (pufferfish and porcupinefish), box-shaped like the boxfish, flattened and compressed (angelfish, butterflyfish, triggerfish and surgeonfish) or depressed (manta, eagle rays and stingrays). Every shape has a specific relationship with the type of environment in which each species lives. Sometimes, the link between habitat and fish is very close, as is the case of some tropical damselfish, which live among the branches of the hard coral; they hide in them in case of danger, and do not leave them even if the coral is temporarily removed from the water. However, fish and coral are just some of the many life forms to be found on the reef, which are numerically dominated by the invertebrates. Molluscs and echinoderms are very common. The former include numerous small sea slugs such as the brightly colored *Phyllidia* or the pale *Chromodoridae*, which crawl over the coral in search of the hydroids and sponges they feed on. During the day cowries, whose shiny shells are covered and protected by a large camouflage coat, can be seen among the crevices in the corals. The best time to observe them is in the early morning, when the molluscs have not yet reached their hiding places and can be photographed and observed unhurriedly. In tourist resorts collecting them is prohibited, as is removing any kind of organism found under water. The protection of the Maldive seas depends not only on the inhabitants of the archipelago, but also on the numerous tourists who visit it every year. Bivalve molluscs are exemplified by the giant clams *(Tridacna gigas)*, large numbers of which populate the seabeds, and when they open they resemble colored eyes among the coral. Numerous sea urchins and starfish also inhabit the reef. Red and yellow *Fromia* starfish mingle with the blue *Linckia laevigata* and the cushion starfish of the *Culcita* and *Choriaster* genera. The large spiny crown of thorns starfish *(Acanthaster planci)*, with its reddish-blue highlights, can be seen from time to time, intent on devouring the polyps of the hard coral. Brittle stars, with their long, slender arms, move slowly in the crevices of the coral and on the sea fans. Because of their size, they often go unnoticed, unlike the crinoids, whose colorful feathery arms are one of the more typical sights on deep dives. Sea urchins are too well known not to be identified, because of their spines, which may be long and slender as in the case of the nocturnal long spined sea-urchins, or as thick as pencils, as in the case of the pencil sea-urchin. There are also numerous crustaceans. Lobsters, crayfish, crabs and hermit crabs scuttle around on the sea bed, especially at night, acting as efficient refuse disposal teams. Some live in close symbiosis with larger animals (like anemones and fish), undertaking to clean them of parasites and organic debris in exchange for protection and food.

*121 top Sea slugs (Hypselodoris sp.) often live in contact with sponges.*

*121 center A prawn (Periclimenes sp.) is perfectly camouflaged on the tentacle of an alcyonarian.*

*121 bottom Two Christmas tree worms (Spirobranchus giganteus) have grown in contact with a coral which covers their tubes.*

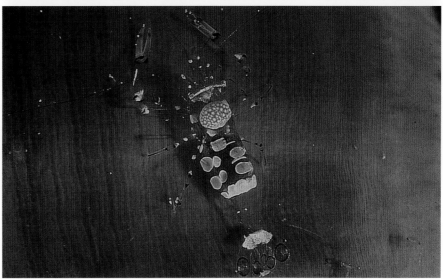

These species, which include the prawns of the *Periclimenes* genus, are usually very small, and it takes a certain amount of experience to see them, but true underwater exploration should include some attention to these and other small inhabitants of the reef. A night dive will generally reveal a more lively and colorful underwater scene: many seemingly dead or unactive creatures spring to life. A night dive in the Maldives can be undertaken in the house reef within close proximity of the resort island.

# THE LAGOONS

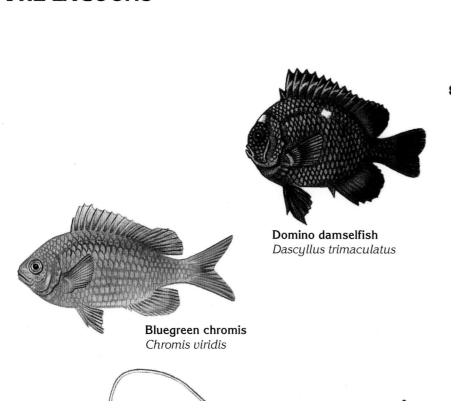

**Domino damselfish**
*Dascyllus trimaculatus*

**Banded dascyllus**
*Dascyllus aruanus*

**Bluegreen chromis**
*Chromis viridis*

**Sixspot goby**
*Valenciennea sexguttata*

**Spotted eagle ray**
*Aetobatus narinari*

**Pinkbar shrimp goby**
*Amblyeleotris aurora*

**Stringray**
*Taeniura melanospilos*

**Halfbeack**
*Hyporhamphus dussumieri*

**Picasso triggerfish**
*Rhinecanthus aculeatus*

122

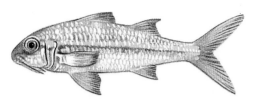

**Yellowfin goatfish**
*Mulloides vanicolensis*

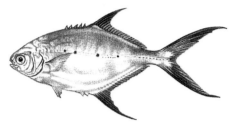

**Smallspotted dart**
*Trachinotus bailloni*

**Blackspot emperor**
*Lethrinus harak*

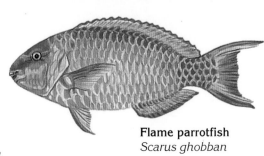

**Sailfin tang**
*Zebrasoma veliferum*

**Flame parrotfish**
*Scarus ghobban*

*122-123 Spotted eagle ray* (Aetobatus narinari).

123

*124-125 Stingray*
(Taeniura
melanospilos).

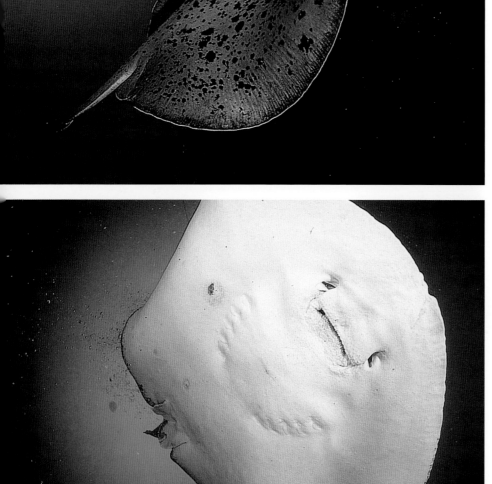

Lagoons are among the best-known and most contrasting seascapes of the Maldives. Their whiteness marks the upper limits of the reef, and the transparent effects created by the sand inside the lagoon seem to be produced by a study in blue. In the lagoons, color is projected outwards, which is perhaps why they look particularly beautiful on an aerial view, whereas from nearby, their main appeal seems to be the clarity of the water. Yet if they are considered more closely, the lagoons, and

especially the *vilus* (the lagoons of the *farus*) will be seen for what they really are: an integral part of the reef and an essential component of the atoll. Generally shallow (5-10 meters deep), the lagoons are the ideal place for snorkelling. If you start from an island beach, the first impression might be that of a desert, but this is because the shallows, which heat up rapidly, are inhospitable to most life forms. In addition, the light reflecting on the coral sand continually produced

by the erosion of dead corals and accumulated in the lagoon by waves and currents makes it difficult to bring the seascape into focus. You will need to go further out in order to see the first fish, like the goatfish, busy probing the sea bed with its long "whiskers" in search of food.

It is even more interesting to go on to places where dark spots can be seen. These spots, which can also be identified from the shore, correspond to blocks of dead coral or to colonies of hard corals, perfectly adapted to life in the lagoon, which have grown on hardened sand or old stretches of hard corals. Some formations bear comparison with a section of true reef in terms of size, although the number of species found will be restricted to the most adaptable. A small sample of the wealth of

life in the surrounding sea can be observed here in perfectly safe waters. As in a model aquarium, the coral is surrounded by clouds of small damselfish, surgeonfish and parrotfish, generally represented by young specimens which can grow without too much difficulty in these protected areas, with relatively few predators. These corals, some of which grow almost to the surface of the water and form miniature lagoons, constitute the ideal stages in our exploration. By swimming from one formation to the next and keeping an eye on the seabed, we may encounter large lurking lizardfish or shy gobies with their camouflage colors, which stand guard in front of their holes, and dart into them if disturbed too much. During exploration of the lagoon you may encounter

126-127 *Yellowfin
goatfish* (Mulloides
vanicolensis).

*127 top Bluescaled
emperor* (Lethrinus
nebulosus).

*127 bottom Flame
parrotfish* (Scarus
ghobban).

currents with a lower temperature; this is a sign which should not be overlooked. These are waters which revitalize the lagoon, entering from the outside through one of the channels interrupting the surrounding reef.

Near these currents hover larger fish, attracted by the flow of food-rich external water, such as the long, tapering halfbeaks, the silvery *Trachinotus* or some Lethrinidae. The edges of the channels are the ideal spot for exploration at sunset or dawn, when eagle rays and stingrays often pass by, on their way to the lagoon for an overnight rest.

*128 top left*
*Domino damselfish*
(Dascyllus
trimaculatus).

*128 bottom left*
*and 129 Bluegreen*
*chromis* (Chromis
viridis).

*128 top right*
*Banded dascyllus*
(Dascyllus aruanus).

*128 bottom right*
*White-belly*
*damselfish*
(Amblyglyphidon
leucogaster).

# THE THILAS

**Silvertip reef shark**
*Carcharinus albimarginatus*

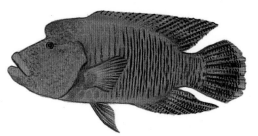

**Great barracuda**
*Sphyraena barracuda*

**Humphead wrasse**
*Cheilinus undulatus*

**Redtooth triggerfish**
*Odonus niger*

**Tuna fish**
*Thunnus alalunga*

**Batfish**
*Platax teira*

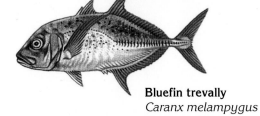

**Bluefin trevally**
*Caranx melampygus*

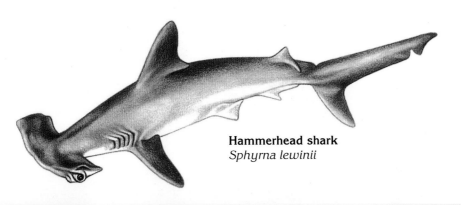

**Hammerhead shark**
*Sphyrna lewinii*

*130-131 Oceanic whitetip shark (Carcharhinus longimanus).*

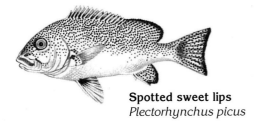

**Spotted sweet lips**
*Plectorhynchus picus*

A pale spot which suddenly appears in front of the *dhoni*, interrupting the surrounding blue, is a sure sign of a *thila*. The term *thila* in the Maldivian language *Divehi* indicates an isolated reef in the open sea; a coral bank which rises from the sea bed almost to the surface. The main characteristic of the *thila*, more accentuated than elsewhere, is the almost constant presence of currents which beat against the coral tower, influencing its development and moulding its shape in accordance with the predominant direction of the waves, whose incessant action sometimes creates fascinating underwater tunnels. The seabed is teeming with every kind

*132 Batfish*
(Platax teira).

*133 top Barracuda*
(Sphyraena qenie).

*133 center*
*Hammerhead shark*
(Sphyrna lewini).

*133 bottom*
*Great barracuda*
(Sphyraena
barracuda).

of life form; fish which are also quite common in other habitats, such as moray eels, soldierfish, stonefish, scorpion fish and clownfish mix with gorgonians, alcyonarians and hard corals. However, the real dominators of the thila are the large pelagic fish. Attracted by the currents, large tuna fish and jacks converge here, chasing shoals of clupeids. Silvertip reef sharks *(Carcharhinus albimarginatus)* and hammerheads *(Sphyrna lewinii)* are by no means rare at some times of year, especially near the *thilas* less commonly frequented by man. More frequently encountered are other sharks such as the grey reef shark, after which the "shark thilas" are named. The range of large fish is completed by the great barracudas which, like the other predators, find a wealth of food in this environment, where muscle power and swimming ability make the difference between life and death. However, as already mentioned, even the waters nearest the *thila* corals hold some surprises, such as an encounter with the large, curious Humphead wrasse or shoals of *Platax*; these may mingle with the redtooth triggerfish *(Odonus niger)* in search of plankton, which soon hide in crevices in the coral or alcyonarians, disturbing a group of placid grunts and leaving only the tips of their tailfins showing.

*134 top Humphead wrasse* (Cheilinus undulatus).

*135 Bluefin trevally* (Caranx melampygus).

*134 bottom Tunafish* (Thunnus alalunga).

# THE PASSES

**Wahoo**
*Acanthocybium solandri*

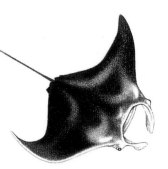

**Giant manta**
*Manta birostris*

**Two-spot red snapper**
*Lutjanus bohar*

**Humpback red snapper**
*Lutjanus gibbus*

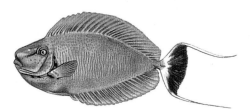

**Vlaming's unicornfish**
*Naso vlammingi*

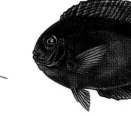

**Chocolate surgeonfish**
*Acanthurus thompsoni*

**Blue-lined surgeonfish**
*Acanthurus lineatus*

**Yellowfin fusilier**
*Caesio xanthonota*

**Bannerfish**
*Heniochus diphreutes*

**Bigeye trevally**
*Caranx sexfasciatus*

*136-137 Bigeye trevally (Caranx sexfasciatus).*

136

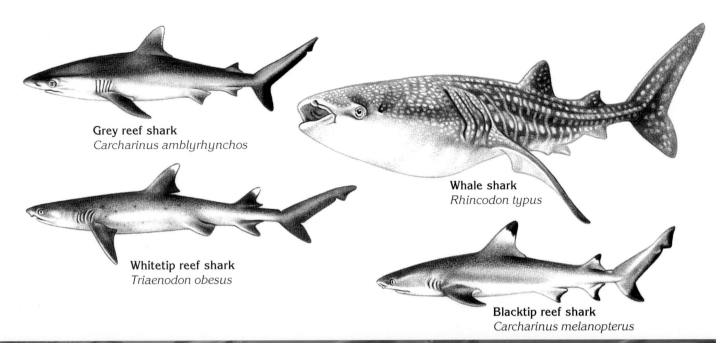

**Grey reef shark**
*Carcharinus amblyrhynchos*

**Whale shark**
*Rhincodon typus*

**Whitetip reef shark**
*Triaenodon obesus*

**Blacktip reef shark**
*Carcharinus melanopterus*

The reefs of the Maldives are surrounded by passes; these are channels which may be over 50 meters deep, with very strong currents produced by the tides which allow divers to cover hundreds of meters effortlessly, as if flying. Diving in these seas requires constant attention, but a good guide, close contact with the support boat and a frequent look at the instruments will allow the diver to observe in safety the outline of grey sharks and blacktip sharks standing out against the blue sea, together with whitetip reef sharks, recognisable by the white tips of the dorsal and tail fins. An encounter with sharks, called *miyaru* by the Maldivians, is almost guaranteed, although despite their reputation, these fish are generally

shy unless they have been encouraged to approach divers by offers of food. More than 20 species of sharks has been identified in the Maldives. In the Maldives there are special diving spots where sharks are common; one of them is the hammerhead shark watching spot near Rasdhu, in Ari Atoll. It is important to remember that shark attacks on snorkellers and divers in the Maldives are unknown. A frequent look at the open sea and the surface will often be rewarded by the sight of a passing shoal of eagle rays, recognisable by their long thin tails and white-spotted backs, or a giant manta, with its white belly and wide wings. Far more unusual, but not unlikely, is an encounter with the whale shark.

*138 top and 138-139
Whale shark
(Rhincodon typus).*

*138 bottom and 139 top
Giant manta (Manta
birostris).*

*141 top and bottom
Blacktip reef shark
(Carcharhinus
melanopterus).*

*140-141 and 141 center
Grey reef shark
(Carcharhinus
amblyrhynchos).*

*140 bottom left
Silvertip shark
(Carcharhinus
albimarginatus).*

*140 bottom right
Whitetip reef shark
(Triaenodon
obesus).*

141

*142 top Bannerfish*
(Heniochus
diphreutes).

*142 center Yellowfin
fusiliers* (Caesio
xanthonota).

*142 bottom
Vlaming's
unicornfish*
(Naso vlamingii).

*142-143 Neon
fusiliers*
(Pterocaesio tile).

The waters of the pass are also regularly visited by wahoos, elongated fish with a similar shape to the barracuda, which swim so fast as to be almost invisible. Large fish are not characteristic of the open sea, however. Near the flattish beds of the canyon-like channels dominated by long sea whip corals, black corals and rigid colonies of *Tubastreae*, swim shoals of jacks mingling with the unmistakeable shapes of *Naso vlamingii* and other surgeonfish, and yellow and blue clouds of fusiliers *(Caesio xanthonota)*. The latter, which indicate the presence of currents and clouds of plankton, constitute one of the basic links in the many food chains developed in the reef, the last links in which include some of the larger snappers.

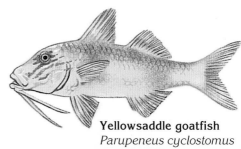

**Yellowsaddle goatfish**
*Parupeneus cyclostomus*

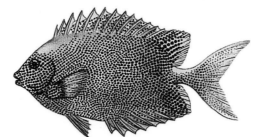

**Stellate rabbitfish**
*Siganus stellatus*

**Lizardfish**
*Synodus variegatus*

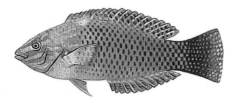

**Checkerboard wrasse**
*Halichoeres hortulanus*

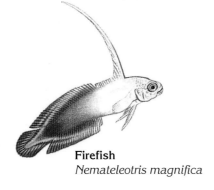

**Firefish**
*Nemateleotris magnifica*

**Black spotted gardeneel**
*Heteroconger hassi*

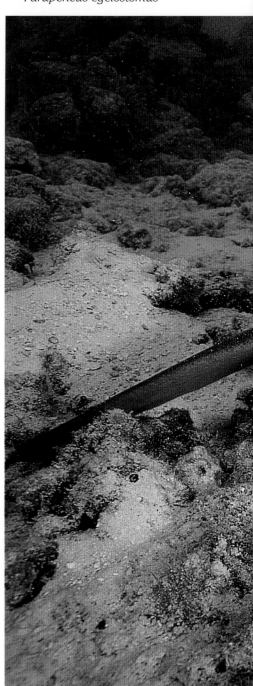

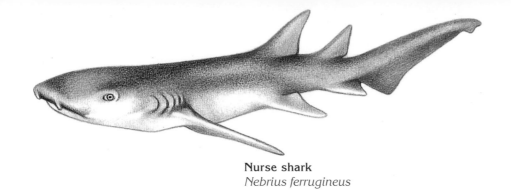

**Nurse shark**
*Nebrius ferrugineus*

*144-145 Nurse shark (Nebrius ferrugineus).*

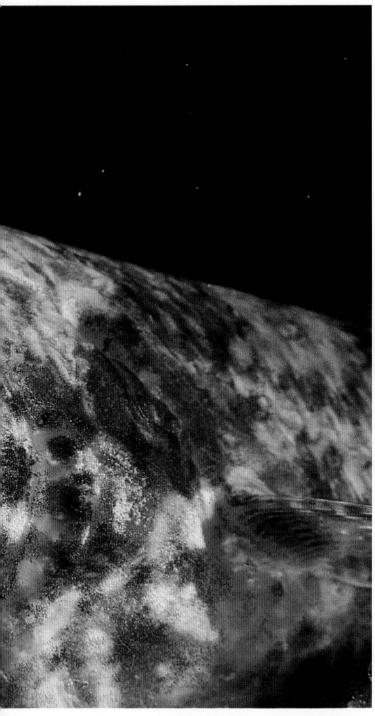

*146-147 Lizardfish
(Synodus
variegatus).*

*146 bottom left
Red firefish
(Nemateleotris
magnifica).*

*146 bottom right
Purple firefish
(Nemateleotris
decora).*

*147 Black spotted
gardeneel
(Heteroconger
hassi).*

The sandbeds which interrupt the reef closest to
the surface at irregular intervals, and become
almost predominant alongside the flat steps
which break up the outer slope of the reef,
appear to be devoid of life. This may be due
to the lack of attention given to these beds
by divers, justifiably distracted by the
multicolored reef fish, but also because
they often empty at the approach of man.
This happens, for example, whenever divers
approach the gardeneels *(Heteroconger hassi)*;
the eels rapidly take refuge in holes dug in the
sand, a habit they share with the fire dartfish,
which live in pairs in sandy clearings at the
foot of the reef or at the entrance to small grottos.
The sand which covers the floor of the larger
grottos and the seabed from which umbrellas

of acropore grow become the favourite hiding places of nurse sharks during the day. If the grottos are carefully examined, the diver is quite likely to catch a glimpse of these sharks, which are creatures of habit and quite happy to be approached. However, it is best never to move between the shark and the exit, as the animal, irritated by the presence of the diver, might make a rapid exit, knocking aside anything in its way. The sand is not only a convenient hiding place. For many fish, like wrasses, mullets and rabbitfish, it is a meeting point or a source of food, represented by worms, crustaceans, molluscs and echinoderms.

*148 top  Cheeklined splendour wrasse* (Oxycheilinus diagrammus).

*148 center Longbarbel goatfish* (Parupeneus macronema).

*148 bottom Checkerboard wrasse* (Halichoeres hortulanus).

*148-149 and 149 top Yellowsaddle goatfish* (Parupeneus cyclostomus).

**Blotcheye soldierfish**
*Myripristis murdjan*

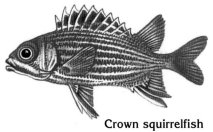

**Crown squirrelfish**
*Sargocentron diadema*

**Sabre squirrelfish**
*Sargocentron spiniferum*

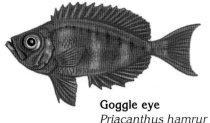

**Goggle eye**
*Priacanthus hamrur*

**Oriental sweetlips**
*Plectorhynchus orientalis*

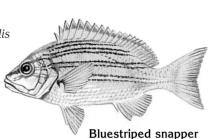

**Bluestriped snapper**
*Lutjanus kasmira*

**Giant moray**
*Gymnothorax javanicus*

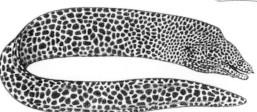

**Honeycomb muray**
*Gymnothorax favagineus*

**Leopard moray**
*Gymnothorax undulatus*

*150-151 Giant moray (Gymnothorax javanicus).*

Corals grow in accordance with geometrical
patterns which vary from species to species.
Some hard corals are always solid, others are
branched, and others again adapt to the prevalent
conditions of light and hydrodynamics at the point
in which they settled at the larval stage.
The continual growth of the corals is matched by
the gradual death of the polyps in the lower
layers. These polyps are transformed into
calcareous rock, which can be attacked by
encrusting and perforating organisms, or by
chemical agents which undermine its
compactness. In this way the reef is radically
transformed; a myriad of cracks and grottos

*152-153 and 153 top Bluestriped snappers (Lutjanus kasmira).*

*153 center Oriental sweetlips (Plectorhynchus orientalis).*

*153 bottom School of glassfish (Parapriacanthus guentheri).*

form in which numerous species of fish take refuge or establish their territory. Moray eels are commonly encountered in these environments. The various species, including the large *Gymnothorax javanicus* and the black and white leopard moray, all behave in the same way, poking their heads and a small part of their bodies out of their holes. Their terrifying appearance is misleading. The wide-open mouth merely enables them to breathe better, and is not a threat. Their sedentary habits during the day (as opposed to the night, when they turn into fierce predators) mean that they can be closely observed, once the awe they inspire has been overcome, to discover some interesting aspects of their lives, like the grooming prawns which

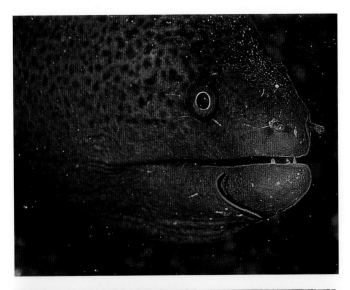

154  Yellow-edged moray (Gymnothorax flavimarginatus).

155 top left Ribbon eel quaesita (Rhinomuraena quaesita).

155 bottom left Honey combmoray (Gymnothorax favagineus).

155 top right Giant moray (Gymnothorax javanicus).

155 bottom right Longtail moray (Strophidon sathete).

often surround their snout and mouth. The red soldierfish and the squirrelfish, whose large eyes, typical of *Priacanthus hamrur*, immediately identify them as nocturnal inhabitants of the reef, also take refuge in grottos. During the daytime these fish remain in shadow; they form shoals, sometimes mixed ones, in the darkest parts of the reef, leaving the twilight areas to the more colorful oriental sweetlips and striped snappers.

*156 top left Whitetipped soldierfish* (Myripristis vittata).

*156 center and bottom left Goggle eye* (Priacanthus hamrur).

*156 right Sabre squirrelfish* (Sargocentron spiniferum).

*157 Goggle eye* (Priacanthus hamrur).

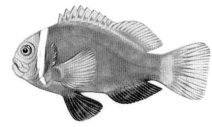

**Half-and-half chromis**
*Chromis dimidiata*

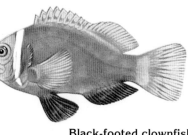

**Black-footed clownfish**
*Amphiprion nigripes*

**Clark's anemonefish**
*Amphiprion clarkii*

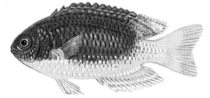

**Caerulean damselfish**
*Pomacentrus caeruleus*

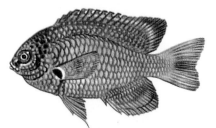

**Indian damselfish**
*Pomacentrus indicus*

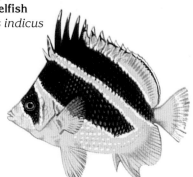

**Indian butterflyfish**
*Chaetodon mitratus*

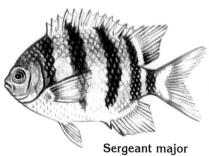

**Sergeant major**
*Abudefduf saxatilis*

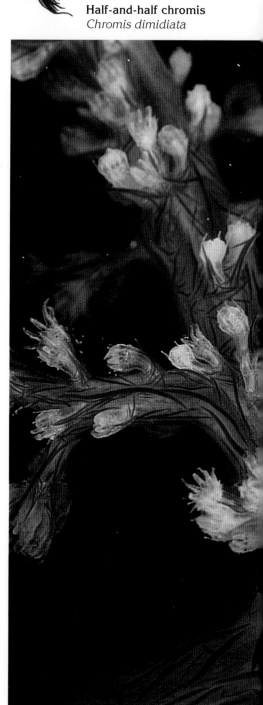

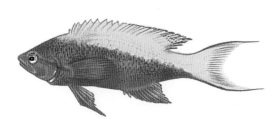

**Yellowtail fairy-basslet**
*Pseudanthias evansi*

**Scalefin anthias**
*Pseudanthias squamipinnis*

*158-159 Goby*
(Bryaninops sp.).

Observing life in the reef, it might almost be imagined that the corals (animals which, with very rare exceptions, are condemned to perpetual immobility by the rock they themselves produce) have transferred their mobility to the fish that surround them. The coming and going of the anthias, continually moving above the reef, indicates the motion of the waves and the approach of larger fish. Their pink, orange and violet clouds form and disappear almost rhythmically, while maintaining a strict ranking, with dominant males and females controlling the territory. More territorial than most are the clownfish, which live, in pairs or small families, inextricably associated with the anemones, into whose tentacles they dive at the first sign of danger, except during the mating season, when the need to defend their eggs makes them aggressive towards all, including man.

While the anthias colonise vast areas of the reef, some Pomacentridae, the colorful damselfish, are closely associated with branched corals such as the acropores or pocillopores. Each shoal lives in its own coral tree, swimming nimbly among the coral during the day and transforming the intertwining branches into a safe haven at night. The *Abudefduf*, with their typical black-striped yellow and white coloring, which live in large shoals, are freer, more independent, and sometimes almost aggressive. They are usually among the first fish encountered by divers, as they remain in the upper layers of the water; they take little notice of humans who cross their pastures. Much deeper, and usually associated with the sea fans and black coral that grow on the steepest slopes of the reef, we may find pairs of Indian butterflyfish, recognisable by their broad diagonal black stripes and the strange hump on their backs.

*160-161 and 161 center and bottom Black-footed clowinfish* (Amphiprion nigripes).

*161 top Clark's anemonefish* (Amphiprion clarkii).

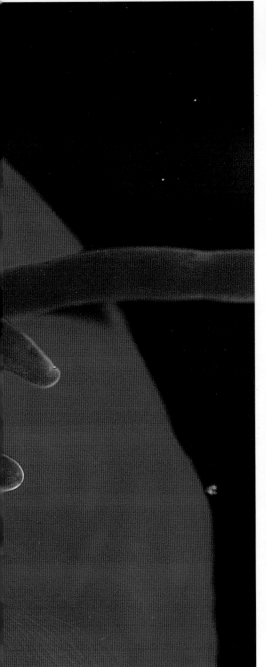

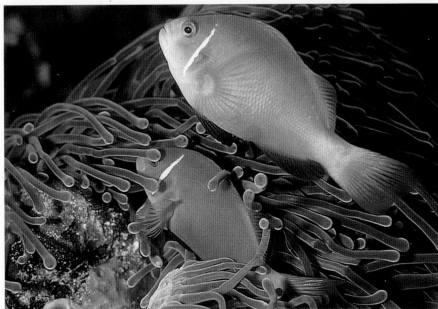

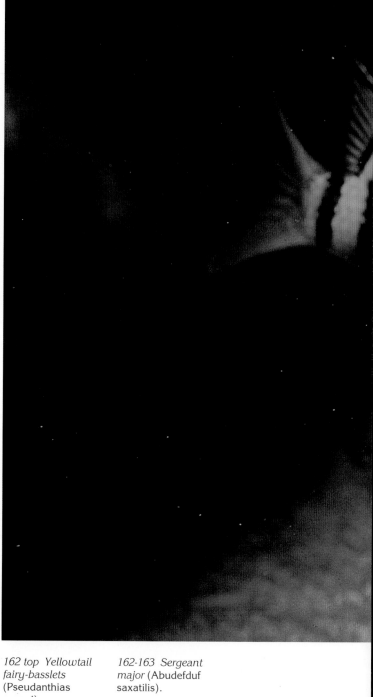

*162 top Yellowtail fairy-basslets* (Pseudanthias evansi).

*162 center and bottom Scalefin anthias* (Pseudanthias squamipinnis).

*162-163 Sergeant major* (Abudefduf saxatilis).

*163 bottom Half-and-half chromis* (Chromis dimidiata).

164 top  Goby
(Bryaninops sp.).

164-165  Linear
blenny (Ecsenius
lineautus).

165 top  Blenny
(Ecsenius sp.).

165 center  Goby
(Bryaninops sp.).

165 bottom
Cardinalfish
(Apogon sp.).

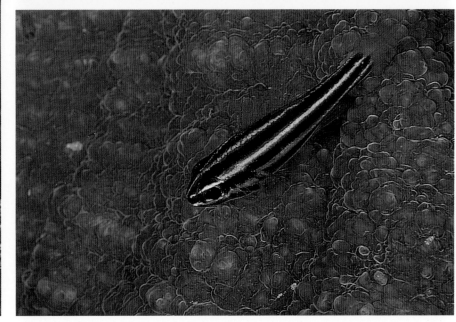

**Bullethead parrotfish**
*Scarus sordidus*

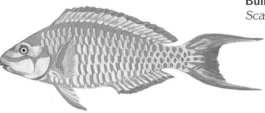

**Longnose parrotfish**
*Hipposcarus harid*

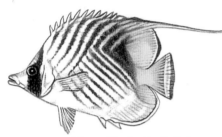

**Threadfin butterflyfish**
*Chaetodon auriga*

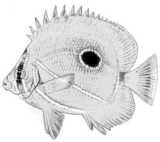

**Bennett' butterflyfish**
*Chaetodon bennetti*

**Racoon butterflyfish**
*Chaetodon lunula*

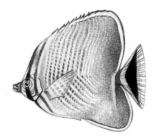

**Triangular butterflyfish**
*Chaetodon triangulum*

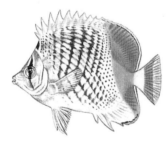

**Madagascar butterflyfish**
*Chaetodon madagascariensis*

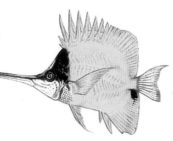

**Long-nosed butterflyfish**
*Forcipiger longirostris*

**Black-backed butterflyfish**
*Chaetodon melannotus*

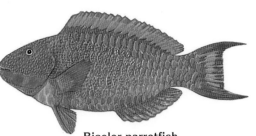

**Bicolor parrotfish**
*Cetoscarus bicolor*

**Steephead parrotfish**
*Scarus gibbus*

*166-167 Racoon butterflyfish* (Chaetodon lunula).

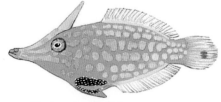

**Harlequin pilefish**
*Oxymonacanthus longirostris*

*168-169 Red-tailed*
*butterflyfish*
(Chaetodon
collare).

*168 top Triangular*
*butterflyfish*
(Chaetodon
triangulum).

*168 center*
*Threadfish*
*butterflyfish*
(Chaetodon
auriga).

*168 bottom*
*Yellowhead*
*butterflyfish*
(Chaetodon
xanthocephalus).

When observing the reef, it is not uncommon to catch a glimpse among the coral of butterflyfish, all of which have flattened bodies that allow these species to move nimbly among the coral. This is the result of specific adaptation to their main food resource, represented in the case of most Chaetodontidae by small coral polyps. In this case the hard corals become not only a place to hide, but also a basic food resource. If you follow a butterflyfish under water, you will see it approach the hard corals and strike its surface with rapid blows of the mouth, as if pecking it. In this way, the fish brings its tiny pincer-like teeth into operation, using them to grip the invisible polyps protruding from the coral. The evolution of this family has produced a wide variety of colors that allow each species to recognize one another and, therefore, defend the territories where they obtain their food. Sometimes, the differentiation also extends to the mouth, as in the case of the long-nosed butterflyfish *Forcipiger longirostris*, whose long snout is ideal for capturing the organisms that live in the innermost depths of the hard corals. Parrotfish are also specialized coral predators; their teeth are fused to form four plates (two upper and two lower plates) which constitute such a strong beak that they can make deep grooves in the surface of the coral, leaving easily visible signs, especially in the globular or cerebral hard corals. At regular intervals the parrotfish will be seen to be surrounded by a white cloud, which is soon deposited on the sea bed after attracting small wrasses and other fish. The cloud actually consists of the undigested remains of coral, which is transformed into fine coral sand.

170-171 Lined
butterflyfish
(Chaetodon
lineolatus) *with
giant moray*
(Gymnothorax
javanicus).

170 bottom left
Redfin butterflyfish
(Chaetodon
trifasciatus).

170 bottom right
Long-nosed
butterflyfish
(Forcipiger
longirostris).

171 top Black-
backed butterflyfish
(Chaetodon
melannotus).

171 bottom
Harlequin filefish
(Oxymonacanthus
longirostris).

172-173 Indian
longnose parrotfish
(Hipposcarus
harid).

172 bottom
Indian sheephead
parrotfish (Scarus
strongylocephalus).

173 top Swarthy
parrotfish (Scarus
niger).

173 center
Bullhead parrotfish
(Scarus sordidus).

173 bottom
Parrotfish
(Scaridae).

175 *Tassled scorpionfish* (Scorpaenopsis oxycephalus).

**Schultz's pipefish**
*Corythoichthys schultzi*

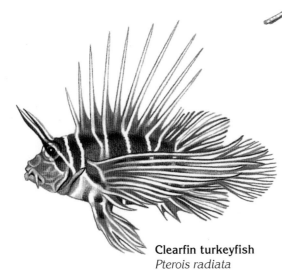

**Clearfin turkeyfish**
*Pterois radiata*

**Leaf scorpionfish**
*Taenionotus triacanthus*

**Turkeyfish**
*Pterois volitans*

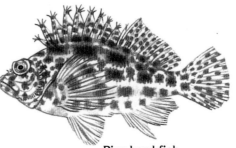

**Pixy hawkfish**
*Cirrhitichthys oxycephalus*

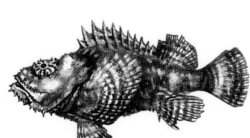

**Devil scorpionfish**
*Scorpaenopsis diabolus*

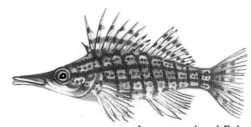

**Longnose hawkfish**
*Oxycirrhithes typus*

176-177 *Forster's hawkfish* (Paracirrhites forsteri).

*176 bottom Pixy hawkfish* (Cirrhichthys oxycephalus).

If you swim close to the coral, but far enough away to ensure that it is not damaged by an accidental blow with a flipper, you will see a range of fish that are not always immediately recognisable from a distance by their color or size. These include the slender pipefish which, with their elongated shape, seem to crawl rather than swim among the coral, hiding in every nook and cranny. Equally well camouflaged is the *Oxycirrhites typus*, whose bright red and white checked pattern is ideal for concealing the fish among the branches of the gorgonians where it likes to live. As if its misleading color were not enough, the longnose hawkfish, with its long snout, moves rapidly from one side of the gorgonian sea fans to another to escape the diver's curiosity. Another hawkfish, *Cirrhitichthys oxycephalus*, is also well suited to its habitat; it prefers to stay above the hard corals, especially in the acropore umbrellas, where it keeps perfectly still, lying in wait for small prey; it darts out to catch them, taking them by surprise, then returns to its look-out post. Other species typically suited to lie in wait for their prey as a result of natural selection

*177 top Longnose hawkfish* (Oxycirrhites typus).

*177 bottom Monocle hawkfish* (Paracirrhites arcuatus).

are the leaf scorpionfish and devil scorpionfish. The former is named after its flattened body, which allows it to be carried by the currents as if it were a piece of seaweed torn from the sea bed, and then approach the small fish it feeds on with slight movements of its fins. The devil scorpionfish, dangerous because of the venomous spines on its dorsal fin, retains the typical habits of the scorpionfish, which prefer to stay on the sea bed, exploiting their color and the protuberances on their bodies to resemble a piece of rock covered with seaweed and other encrusting organisms.
The turkeyfish, of the *Pterois* genus, are prettier and more elegant; they are armed with fearful venomous spines and long, wing-like fins with which they push small fish towards the reef wall, closing them in a trap from which there is no escape.

*178-179 Leaf scorpionfish* (Taenianotus triacanthus).

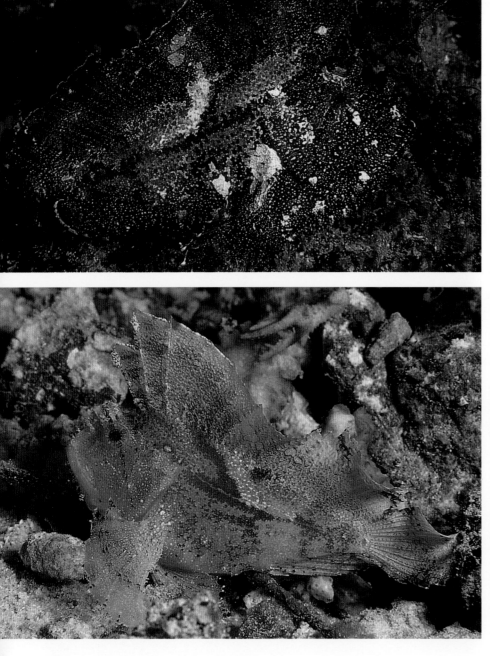

*180 Tassled scorpionfish* (Scorpaenopsis oxycephalus).

*181 left* Turkeyfish (Pterois radiata).

*181 right top* Turkeyfish (Pterois volitans).

*181 right center* Raggy scorpionfish (Scorpaenopsis venosa).

*181 right bottom* Tassled scorpionfish (Scorpaenopsis oxycephalus).

181

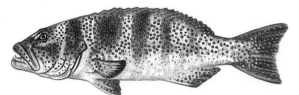

**Black-saddled coraltrout**
*Plectropomus laevis*

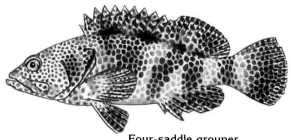

**Four-saddle grouper**
*Epinephelus spilotoceps*

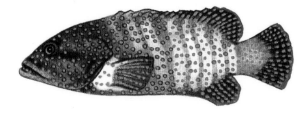

**Peacock grouper**
*Cephalopholis argus*

**Coral grouper**
*Cephalopolis miniata*

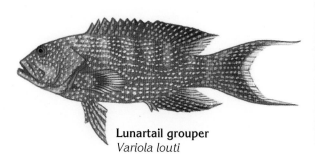

**Lunartail grouper**
*Variola louti*

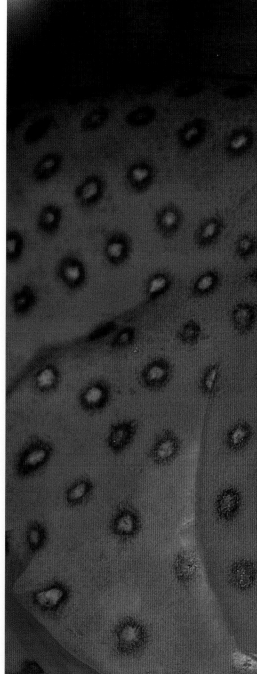

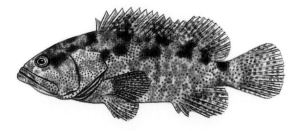

**Smalltooth grouper**
*Epinephelus microdon*

**Redmouth grouper**
*Aethaloperca rogaa*

*182-183  Coral grouper
(Cephalopholis
miniata).*

*184 left top and center Greasy grouper (Epinephelus tauvina).*

*184 left bottom White lined grouper (Anyperodon leucogrammicus).*

*184 right Peacock grouper (Cephalopholis argus).*

*185 Black-saddled coral trout (Plectropomus laevis).*

Among the most common fish on the coral reef, found from just a few meters down to beyond the safety limits for divers, are the groupers, called *faana* by the Maldivians. Their large mouth, big tail fin and massive but elegant body make these fish almost unmistakeable. When swimming along the reef you will soon see that the groupers are the dominant species in this environment.

They can be seen, alone or in pairs, inside grottos or outside, gliding among the hard corals with no apparent effort, intent on patrolling their territory. It isn't easy to follow them.

As soon as they realize they are being tailed, they accelerate or hide under a coral mushroom or in a crevice whose presence we had never suspected; however, it is sure to be well known to the fish, which probably has an emergency exit too.

Some species are more frequently seen by day, others at twilight, and others at night.

They are all carnivorous, and prey on fish, octopus and crustaceans.

As they grow, groupers change gender from female to male, which is why the largest specimens are always males. It is not unusual for the change of gender to be followed by a change in coloring, as in the case of the coral grouper *(Cephalopholis miniata)*; the younger specimens present a yellow-orange color with pale blue spots, while the adults are red with dark blue spots. Other species may change color during the day, as in the case of *Variola louti*, easily recognisable by its crescent-shaped tail, which loses its bright red color at night, taking on a brownish shade for camouflage purposes.

*186-187 and 186 top*
Potato cod
(Epinephelus
tukula).

*186 left top  Lunartail*
*grouper* (Variola
louti).

*186 left bottom*
*Coral grouper*
(Cephalopholis
miniata).

*187 top*
*Peacock grouper*
(Cephalopholis
argus).

**Redbreasted Maori wrasse**
*Cheilinus fasciatus*

**Axilspot hogfish**
*Bodianus axillaris*

**Cleaner wrasse**
*Labroides dimidiatus*

**Bird wrasse**
*Gomphosus caeruleus*

**Black pyramid butterflyfish**
*Hemitaurichthys zoster*

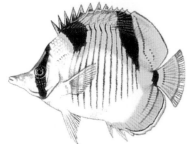

**Masked bannerfish**
*Heniochus monoceros*

**Double-saddled butterflyfish**
*Chaetodon falcula*

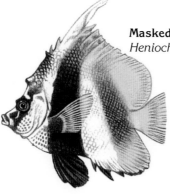

**Blue-faced angelfish**
*Pomacanthus xanthometopon*

**Moon wrasse**
*Thalassoma lunare*

**Trumpetfish**
*Aulostoma chinensis*

**Yellowtail wrasse**
*Anampses meleagrides*

**Cornetfish**
*Fistularia commersonii*

*188-189 Double-saddled butterflyfish (Chaetodon falcula).*

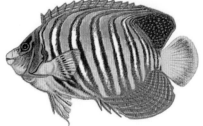

**Royal angelfish**
*Pygoplites diacanthus*

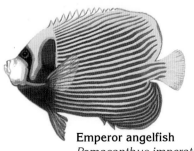

**Emperor angelfish**
*Pomacanthus imperator*

189

In view of the way in which coral grows, the variety of species found in the various parts of the reef, and the difficulty of discerning a pattern in these underwater gardens, the comparison with a jungle is obvious. New halls and new marvels appear round every corner, and behind a hard coral formation or pinnacle an infinite number of others can be glimpsed. The key to understanding might be represented by the fish, but even in their world, shapes and colors seem designed to cause confusion. Only after several dives will you begin to notice some repetition in lifestyles and in the associations between fish and coral, although these repetitions can never dispel our amazement.

*190-191 Emperor
angelfish, adult
(Pomacanthus
imperator).*

*191 center Blue-
faced angelfish
(Pomacanthus
xanthometopon).*

*191 top Royal
angelfish
(Pygoplites
diacanthus).*

*191 bottom Emperor
angelfish, juvenile
(Pomacanthus
imperator).*

The standard features of the reef include "cleaning" stations, where grooming fish are almost always to be found. Fish wishing to be cleaned approach the "station" and display a behaviour pattern which clearly demonstrates their intentions. The cleaner wrasse *(Labroides dimidiatus)* answer these signals; they approach the "customer" and carefully inspect its skin, gills and teeth, eliminating parasites and food residues. All fish use these stations, and a diver who keeps quite still may find himself being inspected by a particularly enterprising cleaner fish. Cornetfish and trumpetfish swim among the gorgonian sea fans in search of prey, or a large wrasse or parrotfish to hide behind. The infinite variety of Labridae swim in an unmistakeable manner among the coral, with rapid strokes of their pectoral fins; they take little notice of man, but concentrate on looking for food or mates, mingling with the butterflyfish and angelfish which brighten up the reef during the day, but disappear at night. At sunset, the Labridae go down to the sea bed, where they sleep buried in the sand or lying on one side in the shelter of a stone.

On the other hand, the butterflyfish and angelfish take refuge in crevices at various depths, and often change color to dull shades, quite different from the bright ones displayed during the day.

*192-193  Masked bannerfish* (Heniochus monoceros).

*193 top  Longfin bannerfish* (Heniochus acuminatus).

*193 center Trumpetfish* (Aulostoma chinensis).

*193 bottom Cornetfish* (Fistularia commersonii).

# HORNS, SPINES AND TEETH

**Dussumier's surgeonfish**
*Acanthurus dussumieri*

**Convict surgeonfish**
*Acanthurus triostegus*

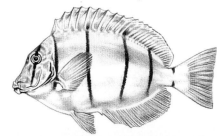

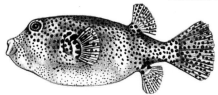

**Moorish idol**
*Zanclus cornutus*

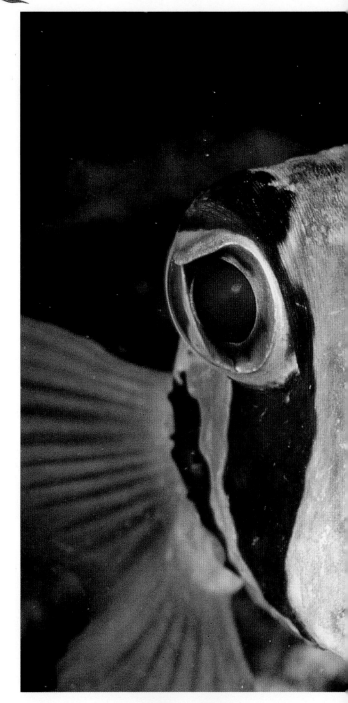

**Blackspotted putterfish**
*Arothron stellatus*

**Bleeker's porcupinefish**
*Diodon liturosus*

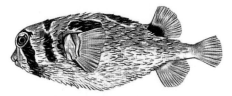

**Orangestriped triggerfish**
*Balistapus undulatus*

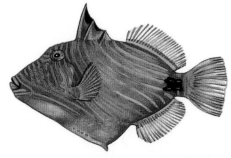

**Boomerang triggerfish**
*Sufflamen bursa*

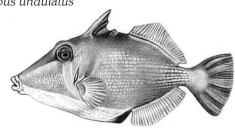

**Titan triggerfish**
*Balistoides viridescens*

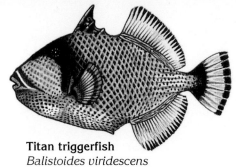

**Blue surgeonfish**
*Acanthurus leucosternon*

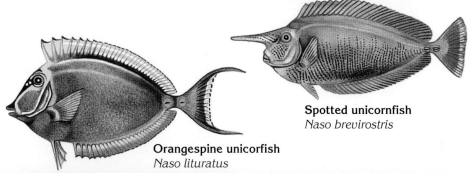

**Orangespine unicorfish**
*Naso lituratus*

**Spotted unicornfish**
*Naso brevirostris*

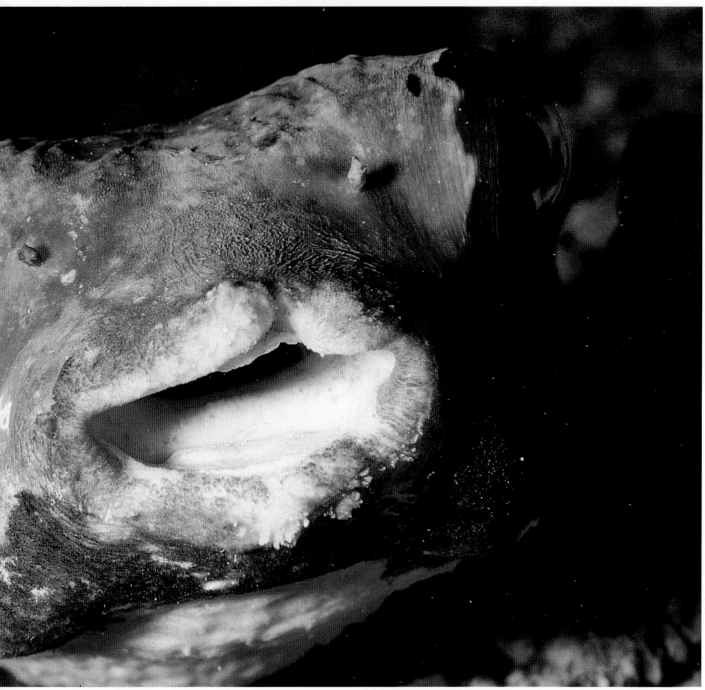

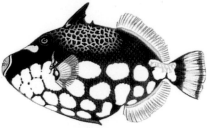

**Clown triggerfish**
*Balistoides conspicillum*

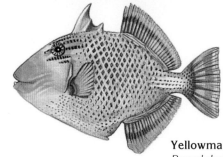

**Yellowmargin triggerfish**
*Pseudobalistes flavimarginatus*

*194-195 Bleeker's porcupinefish (Diodon liturosus).*

*196 Blue surgeonfish (Acanthurus leucosternon).*

*197 top Blue surgeonfish (Acanthurus leucosternon) and blackstreak surgeonfish (Acanthurus nigricauda).*

*197 center left School of elongate surgeonfish (Acanthurus mata).*

*197 center right Spotted unicornfish (Naso brevirostris).*

*197 bottom Spine of unicornfish (Naso lituratus).*

With a few exceptions, fish are usually believed to be harmless. However, they are permanently engaged in a fight for survival, and are equipped with defensive and offensive weapons comparable to those of land animals. Among the best armed are the surgeonfish, named after the sharp spines situated at the sides of the caudal peduncle. If this part of the body is carefully observed, you will see spots of contrasting color which indicate the position of the spines. These may be single retracting spines, as in the case of the *Acanthurus* genus, or fixed double spines as in the case of the *Naso* genus, the unicornfish. These spines are generally used to settle territorial disputes between fish, but it is best to take care when approaching a shoal of surgeonfish, as they might cause an accidental injury.

Triggerfish are armed with strong teeth. These large, agile fish which have an erectile dorsal fin possess teeth that can easily break mollusc shells and the skeletons of sea urchins. During the mating season their territorial instinct is accentuated, as is their aggression, so it is best to beware of the larger specimens, especially the species *Balistoides viridescens*.

*198-199
Orangestriped
triggerfish*
(Balistapus
undulatus).

*199 center
Yellowmargin
triggerfish*
(Pseudobalistes
flavimarginatus).

*199 top  Titan
triggerfish*
(Balistoides
viridiscens).

*199 bottom
Clown triggerfish*
(Balistoides
conspicillum).

The spines of other sea predators, such as the porcupinefish *(Diodon liturosus)*, and those which are displayed when the fish puffs up like the pufferfish *(Arothron stellatus)*, have defensive purposes. Inflation, a characteristic which fascinates divers, is a tiring exercise for both these families, so they should never be touched or goaded to inflate, in order to prevent unnecessary stress and damage to their organs.

*200 top Blackspotted puffer* (Arothron nigropunctatus).

*200 bottom Yellow boxfish* (Ostracion cubicus).

*200-201 Boxfish* (Ostracion sp.).

*201 top left Black-saddled puffer* (Canthigaster valentini).

*201 top right Map puffer* (Arothron mappa).

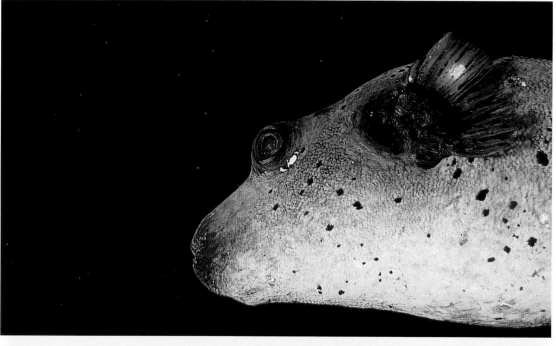

# Malaysia

VIETNAM

THAILAND

Kota Baharu

SOUTH

George Town

Kuala Terengganu

PENINSULAR
MALAYSIA

Ipoh

Kuantan

Kelang • Kuala Lumpur

• Seremban

Melaka •

Johor Bahru

SINGAPORE

N

**Texts**
Andrea and Antonella Ferrari

**Translations**
Studio Traduzioni Vecchia, Milan

# Contents

| INTRODUCTION | page 204 |
|---|---|
| - THE OPEN SEA AND COASTAL WATERS | page 210 |
| - THE CORAL REEF | page 218 |
| - PREDATORS ON THE PROWL | page 236 |
| - BETWEEN LAND AND SEA | page 246 |
| - KNIGHTS IN ARMOUR | page 254 |
| - SANDY AND DETRITAL SEABEDS | page 264 |
| - MOLLUSCS, ECHINODERMS AND OTHER INVERTEBRATES | page 274 |
| - THE ASSOCIATIONS | page 286 |

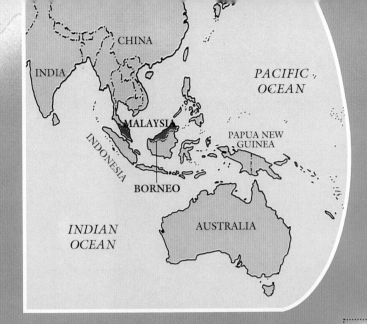

INDIA

CHINA

PACIFIC
OCEAN

MALAYSIA

PAPUA NEW
GUINEA

INDONESIA

BORNEO

INDIAN
OCEAN

AUSTRALIA

# CHINA SEA

*Layang Layang*

*Labuan Island*

## BRUNEI

# MALAYSIA

Miri

Kota Kinabalu

Sandakan

## SABAH

Tawau

## SARAWAK

*Pulau Sipadan*

Sibu

Kuching

# BORNEO

*203 top*
*Malaysian waters it is*
*common to encounter*
*walls of bigeye trevally*
*(Caranx*
*sexfasciatus).*

*203 bottom*
*The bright red*
*of a sea whip of*
*the Elisella genus*
*illuminates the deep*
*water of these seas.*

# INTRODUCTION

## IN THE HEART OF THE INDO PACIFIC

The seabed of peninsular Malaysia and Malaysian Borneo is one of the most spectacular and species rich in the world, thanks to the unique geographic location of the country. Peninsular Malaysia is washed to the west by the Andaman Sea (which in turn borders the Indian Ocean) and by the more southerly Strait of Malacca (which separates it from the

large Indonesian island of Sumatra), and to the east by the southern reaches of the South China Sea. The insular territories of Sabah and Sarawak - which together with Indonesian Kalimantan and the small country of Brunei make up the huge island of Borneo - face the Sulawesi Sea to the east (which separates Borneo from the Indonesian island with the same name), the Sulu Sea to the north (which divides Borneo from the Philippine archipelago) and, finally, the South China Sea (which separates the island of Borneo from peninsular Malaysia to the west). These seas and basins cover an

*204 left  Delicate crinoids and soft corals proliferate thanks to plankton-rich currents.*

*204 top right The Ellisella sea whips offer an ideal feeding spot for plankton feeders such as crinoids.*

*204 bottom right The huge sea fans stretching out from the vertical walls offer infinite inspiration to photographers.*

*205 left  A huge, spectacular example of the Xestospongia testudinaria barrel sponge.*

*205 right  During dives in these waters it is possible to observe dense shoals of passing fish. In the photos: some silver longfin spadefish (Platax teira) top; hundreds of bigeye trevally (Caranx sexfasciatus) center; and sawtooth barracuda (Sphyraena putnamiae), bottom.*

immense area just above the Equator and are characterized by an almost infinite variety of littoral environments, mostly uninhabited, and to date largely unexplored. As yet the area is largely unaffected by the impacts of industrial and agricultural pollution. With careful management this should remain the case for a long time to come. Most coastal fishing is still carried out by local fishermen using traditional methods, even though the deplorable habit of using explosives has already caused serious damage to the reefs at several locations. There are worrying signs that the regular

forays of fleets of fishing boats - especially those from Vietnam and the Philippines, which fish with complete disregard for environmental laws, at times bordering on piracy - are causing impoverishment of pelagic fauna. Serious and perhaps irreparable damage has occurred over the last decade to many local populations of sea creatures: sea moths and seahorses (millions are used by Chinese pharmacopoeia); sea snakes (hundreds of thousands are killed especially in the Philippines for the tanning industry); sea cucumbers (entire populations have been exterminated to supply the

trepang industry, as sea cucumbers are considered a speciality of Asian cooking); sharks of all genera and species (these also, or rather their fins, are unfortunately considered a delicacy by devotees of Chinese cooking); and hawksbill and green turtles (protected in Malaysia but hunted throughout Indonesia for their flesh, eggs and shells). Despite these impacts, the water and seabeds around Malaysia and Indonesia, located in the center of the Indo Pacific basin (the area which stretches for almost thirty thousand kilometers from Madagascar in the west to the easternmost islands of the Pacific), still represent one of the richest marine habitats in the world, with an estimated minimum of more than 3,000 different species of fish. This region, the heart of the genetic wealth of the entire Indo Pacific, has been subjected to the same tropical climate for more than 100 million years, with ideal conditions of light and temperature, which have nurtured the development of a complex differentiation of species. Continental drift, the resulting volcanic activity, and erosion have contributed to the creation of a great variety of ecosystems within which - often even in areas of just a few kilometers - completely different ecological niches exist together in a fascinating evolutive environmental laboratory: unfathomed abyssal depths, rocky basaltic cliffs, shallow sandy or muddy beds, expanses of sea grass, impenetrable mangrove forests and of course spectacular coral reefs (which in Malaysia are mainly fringing reefs). These varied ecosystems represent profoundly different environments whose specific characteristics have led to the evolution of many highly specialized species. The most important geographic region, from a biological point of view, is perhaps the South China Sea basin, about 3,400,000 square kilometers in area, extending from the Equator almost to the Tropic of Cancer and from peninsular Malaysia to the Philippines. The sea depths, taking into account the adjacent areas and satellite basins, vary from just 100 meters (along the continental shelf of the Sonda) to more than 5,000 meters (the abyssal depths of the northern-most part). This sea is greatly affected by the monsoon weather patterns (north-east monsoons from December to February, and the south-west monsoons from June to August) which modify the surface currents and temperatures in shallow water

with predictable regularity. These temperatures generally vary between a minimum of 26°C and a maximum of 30°C. In a tropical regime with more or less constant temperatures such as these, the upward movement of large water masses (which helps to mix the nutrients found at the sea bottom with the water at the top in temperate and cold seas) is not possible. In the tropics the energy supply is ensured by the huge quantities of nutrients flowing into the sea from the fresh water rivers during the monsoons (this is why during the wet season it is common to find turbid water with limited visibility at shallow and

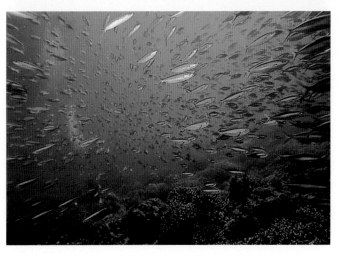

*206 top* These pink anemonefish (Amphiprion perideraion) *live together with several anemones including the pretty giant sea anemone* (Heteractis magnifica).

*206 center* An elegant porcelain crab (Neopetrolisthes maculatus) *peeks out from the sheltering tentacles of the carpet sea anemone* (Stichodactyla mertensii).

*206 bottom Polyps of the cup coral* (Tubastrea sp.), *difficult to see during the day, flaunt their coloured tentacles by night when they open up completely.*

*207 left Elegant feather stars* (Comanthina sp.) *await dusk to capture the plankton on which they feed.*

*207 top right The spotfin lionfish* (Pterois antennata) *are nocturnal and during the day they remain hidden among the corals.*

*207 top center right The green turtles* (Chelonia mydas) *prefer to take refuge in the recesses along the walls.*

*207 bottom center right The docile nurse sharks* (Nebrius ferrugineus) *rest during the day in the shelter of caves and recesses.*

*207 bottom right A shoal of fusiliers* (Caesio sp.) *is a common sight on Malaysian reefs.*

medium depths). The surface currents, high temperatures, and long periods of sunlight also guarantee a high energy supply. These are factors which give rise to the continuum of interconnected ecosystems which exists along the gradual transition from land to oceanic depths. In shallow and medium depths, where there is more sunlight and temperatures are higher, the peak of productivity and biodiversity is reached. The sandy beaches are used as nesting sites by the green turtle (*Chelonia mydas*), the giant leatherback turtle (*Dermochelys coriacea*), and the hawksbill turtle (*Eretmochelys imbricata*),

and as mating sites by striped sea snakes (*Laticauda colubrina*). The mammals, birds, reptiles and shellfish that live in these areas (and their eggs and young) represent an important source of food. Expanses of sea grass offer shelter to many cephalopods (such as squid and octopus) and provide nutrition for green turtles and dugongs, while the hawksbills prefer to eat sponges and soft corals along the reef. The littoral mangrove forests provide food and protection to an enormous number of young creatures gathered together in sorts of nurseries, especially fish, and different types of reptiles and

birds. Other highly specialized species prefer the habitats offered by the sandy and muddy beds, which only appear to be deserted and inhospitable. Many creatures are carried by the currents onto the barrier and fringing reefs, where they reach adulthood, joining the reefs' inhabitants and contributing to the splendour of what has been described as one of the richest ecosystems in the world. Other species, especially large predators such as sharks and tuna, will choose the infinite pathways of the open ocean in search of prey or a companion. However, the conclusion remains the same; however diverse, these

ecosystems are perhaps more inexorably interconnected than any where else on Earth and in each of these habitats - even in the most seemingly inhospitable - there will always be interest for the biologist, naturalist and underwater photographer.

*208 top left*
*This puffer fish (Arothron sp.) presents an unusual indigo colouration.*

*208 center left*
*The splendid bluespotted ray (Tacniura lymma) is easy to observe*

*during the day especially in the sandy areas between the coral formations.*

*208 bottom left*
*This green turtle (Chelonia mydas) is easy to identify with its typical rounded snout.*

*208 right  Many sea whips, in this case a Juncella sp., give shelter to many coral gobies; shown here is a Bryaninops amplus.*

*209 left  Tropical scorpionfish are represented in these waters by many species, most of which belong to the Scorpaenopsis family.*

*209 top right This elegant profile belongs to a Moorish idol (Zanelus cornutus).*

*209 center right Small groups or individual specimens of the lunar-tailed bigeyes (Priacanthus hamrur) often rest in shallow water.*

*209 bottom right The camouflage skills of the frogfish make it difficult to see on the reef.*

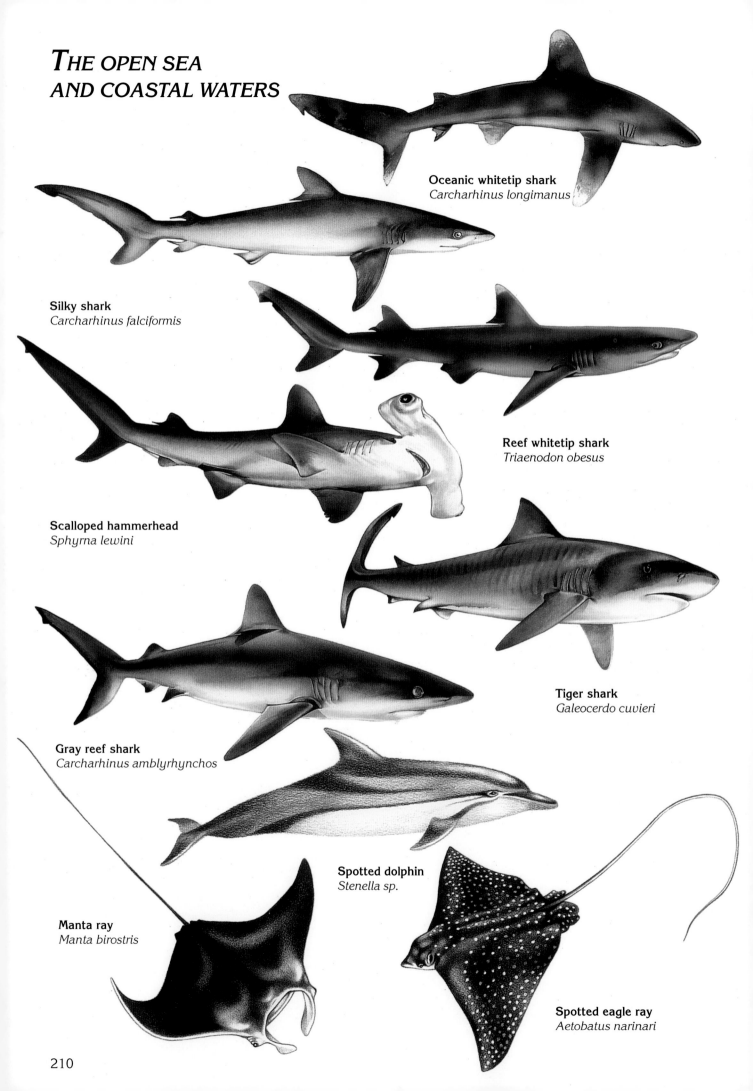

# THE OPEN SEA AND COASTAL WATERS

**Oceanic whitetip shark**
*Carcharhinus longimanus*

**Silky shark**
*Carcharhinus falciformis*

**Reef whitetip shark**
*Triaenodon obesus*

**Scalloped hammerhead**
*Sphyrna lewini*

**Tiger shark**
*Galeocerdo cuvieri*

**Gray reef shark**
*Carcharhinus amblyrhynchos*

**Spotted dolphin**
*Stenella sp.*

**Manta ray**
*Manta birostris*

**Spotted eagle ray**
*Aetobatus narinari*

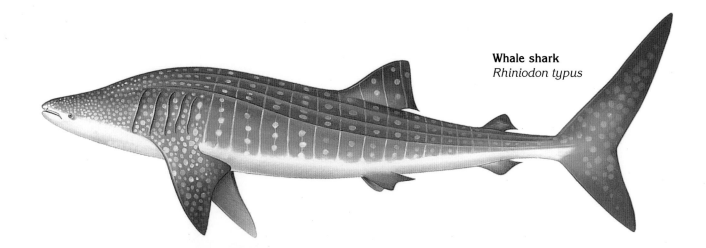

**Whale shark**
*Rhiniodon typus*

*211 Sawtooth barracuda (Sphyraena putnamiae).*

**Sawtooth barracuda**
*Sphyraena putnamiae*

**Bigeye trevally**
*Caranx sexfasciatus*

**Pickhandle barracuda**
*Sphyraena jello*

**Albacore**
*Tunnus alalunga*

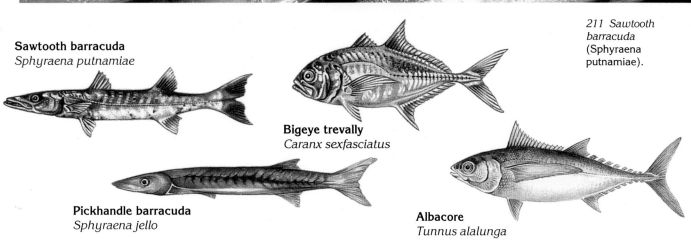

211

*212 top and 212-213*
*Whale shark*
(Rhiniodon typus).

*213 top and bottom*
*Giant manta ray*
(Manta birostris).

The open sea is generally of little interest to amateur scuba divers, who rarely dive there. There are many diving spots in Malaysia - such as Pulau Sipadan or the atoll of Layang Layang - with sea floors right next to the unfathomable depths of the South China Sea or Sulawesi Sea. The open ocean, with its invisible currents and its infinite blue, is at times alarming to the diver and appears to resemble a desert. Yet even this ecosystem, if visited at the right time and in the right area, can offer interesting surprises. This is the kingdom of huge predators and immense shoals of their prey; of the microscopic plankton and the gigantic pelagic filter feeders which feed on this organic "soup". Sometimes these creatures, through choice or need, approach the islands further from the coast, following their prey or obeying the mysterious instinct of reproduction. With much patience and a little luck, it is possible to see species of great interest along the coast of Malaysia during spring, the courting and reproduction season, which coincides with the plankton bloom and the increase in the availability of food. One of the large pelagic filter feeders which lives in this sea is the whale shark *Rhiniodon typus*, the biggest fish in the world, and absolutely harmless, despite its ten meters. Another is the manta ray *Manta birostris*, with its majestic "wing

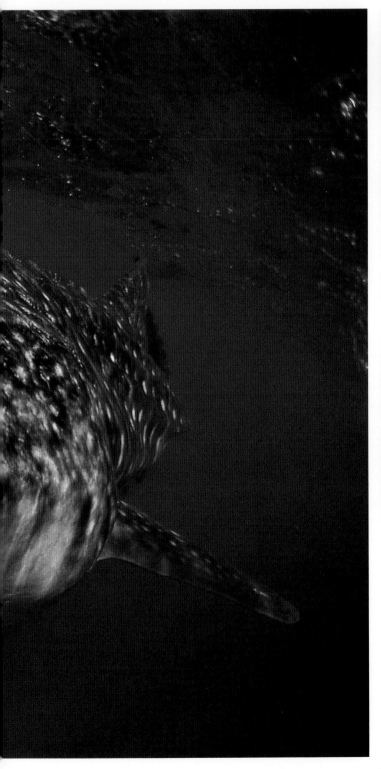

span" which can exceed six meters. These spectacular and extraordinarily elegant creatures, with their graceful and powerful movements, often - if they are not pointlessly harassed - allow divers to swim by their side throughout the dive. The much shyer, but really beautiful sea-eagles, *Aetobatus narinari,* which have an exquisitely white spotted livery, are sometimes encountered in the open sea or along the coast where they search for food.

The oceanic sharks (and those in some way linked to the open water environment), which may be encountered occasionally when diving in Malaysian waters, are: the sericeous sharks *(Carcharhinus falciformis)*; the oceanic white tip shark *(Carcharhinus longimanus)*; the foxshark *(Alopias pelagicus)*, characteristic of deep waters; the silver tip shark *(Carcharhinus albimarginatus)*, generally found in small shoals of females on the deep sandbars; the tiger shark *(Galeocerdo cuvieri)*, more common in sandy, turbid water; and other types of requiem shark. All of these sharks are large, and considered potentially dangerous to man, but it is in fact unusual to encounter them. However, it should be noted

*214 bottom left Reef whitetip shark* (Triaenodon obesus).

*214 bottom right Gray reef shark* (Carcharhinus amblyrhynchos).

*215 center Scalloped hammerhead* (Sphyrna lewini).

*214-215 Oceanic whitetip shark* (Carcharhinus longimanus) *accompanied by the ever-present pilot fish* (Naucrates ductor).

that they are solitary predators whose behaviour changes radically during the night when their habitual daytime caution can give way to surprising aggressiveness. In the waters of Sipadan and Layang Layang during spring, the highly gregarious speckled hammerhead shark *(Sphyma lewini)* can be seen in shoals of some tens to hundreds of creatures. These gatherings appear to be linked to their reproduction rituals. Being extremely shy creatures, however, they generally avoid man, apparently irritated by the hissing of the demand-regulator or the gurgle of the bubbles from the ARA.

Along the coast (at the same time of the year, but sometimes year round) gather gigantic shoals of hundreds to thousands of smaller predators such as the trevally *(Caranx sexfasciatus),* the barracuda *(Sphyraena putnamiae),* and the pickhandle barracuda *(Sphyraena jello).* These genera are represented in the seas by numerous species which are all fairly similar, and have a distinct predatory nature. These gregarious fish can reach extremely high speeds when making their violent surprise attacks. The much larger and impressive dogtooth tuna *(Gymnosarda unicolor)* reaches a length of up to two meters and lives in shoals with a more open structure. These fish unexpectedly arrive from the open sea to explore the reefs further offshore and decimate the shoals of carangids found there. Their rapid and brutal forays are one of the most impressive sights to be seen in these waters. While pelagic sharks and rays are generally characterised by countershading or low visibility camouflage (livery with a bluish or dark grey back and a paler, or even white, belly which makes them difficult to see from either above or below), the bony fish of the open sea,

which generally live in smaller shoals, usually have a silvery, in some cases even chromium-plated look. The reflections from this livery probably helps to temporarily disorient both the predators and their potential prey. Finally, many species of cetaceans are found in Malaysian waters, for example dolphins of the *Stenella* genus, killer whales *(Orcynus orca),* sperm whales *(Physeter catodon)* and various species of finwhales. These are, however, difficult for a diver to approach.

*216 and 217 bottom left Sawtooth barracuda (Sphyraena putnamiae).*

*216-217 and 217 bottom right Bigeye trevally (Caranx sexfasciatus).*

# THE CORAL REEF

**Staghorn coral**
*Acropora palmata*

**Christmas tree worms**
*Spirobranchus giganteus*

**Giant sea anemone**
*Heteractis magnifica*

**Cup coral**
*Tubastrea sp.*

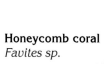

**Honeycomb coral**
*Favites sp.*

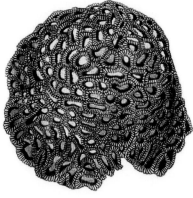

**Mushroom coral**
*Fungia sp.*

**Mushroom anemone**
*Actinodiscus sp.*

**Plate fire-coral**
*Millepora platyphylla*

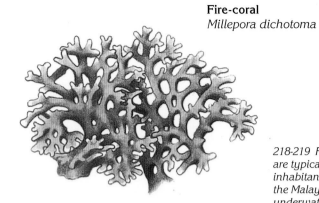

**Fire-coral**
*Millepora dichotoma*

*218-219 Feather stars
are typical
inhabitants of
the Malaysian
underwater world.*

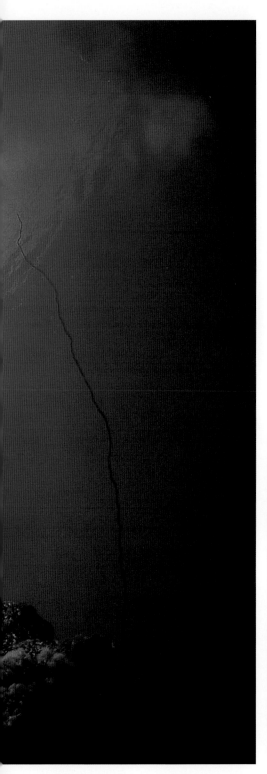

Synonymous with tropical seas and exotic sea life, the reef is believed by many to be the richest ecosystem in the world. The coral reef along the Malaysian coast is one of the most beautiful and least polluted on Earth. Mainly composed of limestone which makes up the exoskeleton of hard corals, and rich in resident or migratory species (a healthy reef can give shelter to more than 3,000 species), the reef is an autonomous ecosystem with a predominantly closed cycle. The three main types of coral reef are all represented in these seas: the littoral reef running more or less parallel to the coast (a type of natural breakwater linked to the fringing reef) the pelagic reef (linked to anomalous elevations of the seabed

*220-221 Leather coral* (Sarcophyton trocheliophorum).

*220 bottom left Staghorn coral* (Acropora sp.) *and* Porites sp..

*220 bottom right False clown anemonefish* (Amphiprion ocellaris) *in their anemone.*

*221 center Soft coral* (Dendronephthya sp.).

whose surface structures may occasionally emerge at low tide like at Sipadan), and the coral atoll (a ring structure with an internal lagoon normally associated with the presence of ancient submerged volcanoes, such as Layang Layang). In any case, the side of the coral reef exposed to the open sea is often affected by strong currents and constant erosion. These areas offer shelter and food sources to biological communities and organisms which are different to those found inside the lagoons.

Some of the most common species responsible for the slow but constant growth of the coral reef are the unmistakable horny formations of the colonies of *Acropora*; the lamellate, yellowish structures of the *Millepora* genus fire coral; the huge domes of *Porites,* often thousands of years old; the lamellate and tabular *Montipora* genus structures; and the extraordinarily flat and broad structures of other species of the *Acropora* genus. This incredibly vast and multiform assemblage of corals is a fragile structure giving shelter to a myriad of fish, molluscs and shellfish. The huge, more flexible sea fans of the *Subergorgia* genus, the dense colonies of black coral of the

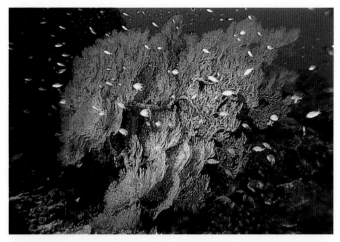

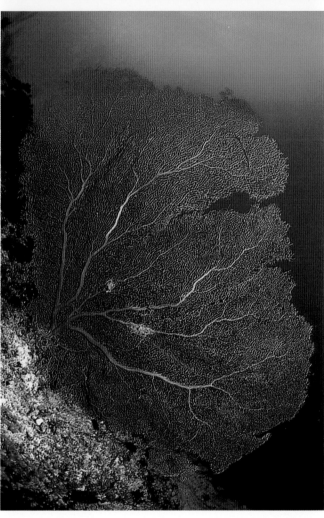

*Anthipathes* genus, and the gigantic barrel sponges *(Xestospongia testudinaria)* are just some of the many sessile organisms whose presence contributes to the creation of the complex and fascinating environment of the deeper reef.

*222 and 223 bottom right* Gorgonian sea fan (Subergorgia sp.).

*223 left* Barrel sponge (Xestospongia testudinaria).

*223 top right* Black coral (Anthipathes sp.).

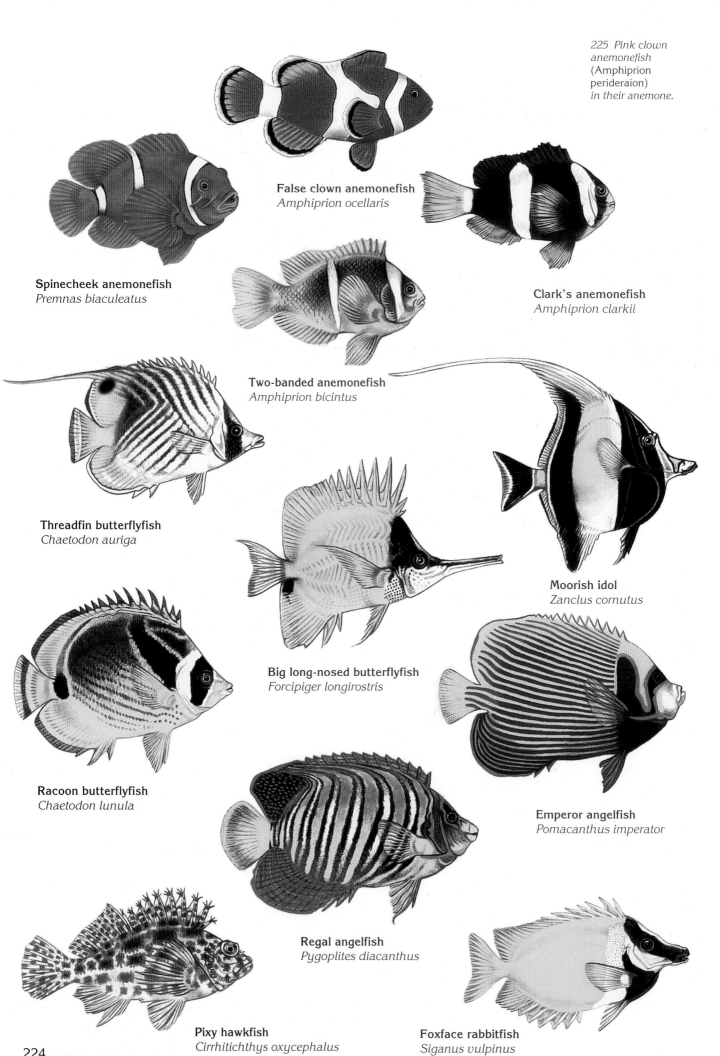

225 Pink clown anemonefish (Amphiprion perideraion) in their anemone.

False clown anemonefish
*Amphiprion ocellaris*

Spinecheek anemonefish
*Premnas biaculeatus*

Clark's anemonefish
*Amphiprion clarkii*

Two-banded anemonefish
*Amphiprion bicintus*

Threadfin butterflyfish
*Chaetodon auriga*

Moorish idol
*Zanclus cornutus*

Big long-nosed butterflyfish
*Forcipiger longirostris*

Racoon butterflyfish
*Chaetodon lunula*

Emperor angelfish
*Pomacanthus imperator*

Regal angelfish
*Pygoplites diacanthus*

Pixy hawkfish
*Cirrhitichthys oxycephalus*

Foxface rabbitfish
*Siganus vulpinus*

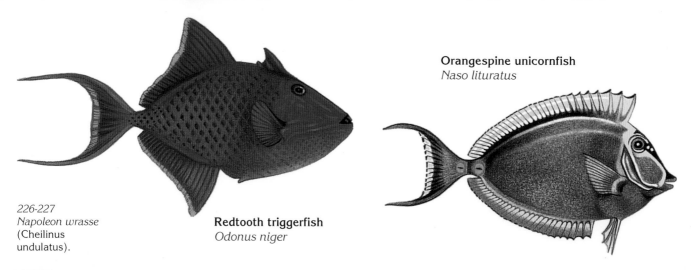

**Orangespine unicornfish**
*Naso lituratus*

*226-227*
*Napoleon wrasse*
(Cheilinus
undulatus).

**Redtooth triggerfish**
*Odonus niger*

**Bignose unicornfish**
*Naso vlammingi*

**Clown triggerfish**
*Balistoides conspicillum*

**Powder-blue
surgeonfish**
*Acanthurus
leucosternon*

**Bullethead parrotfish**
*Scarus sordidus*

**Bumphead
parrotfish**
*Bolbometopon
muricatum*

**Black-saddled toby**
*Canthigaster valentini*

**Scribbled filefish**
*Aluterus scriptus*

**Yellow boxfish**
*Ostracion cubicus*

**Bluestriped fangblenny**
*Plagiotremus rhinorhynchus*

**Longfin
spadefish**
*Platax teira*

227

The coral reef is the kingdom of curiously-shaped
and brightly-colored creatures typically found
in tropical seas, whose adaptation to their
respective evolutive niches has often resulted
in the development of bizarre profiles and
extraordinarily colorful liveries.
The most common inhabitants of the Malaysian
reef include several species of the *Chaetodon*
genus, the butterflyfish (which generally feed on
the polyps of the coral and tiny organisms in
general); the *Amphiprion* genus, the clown
anemone fish (symbionts of the spectacular sea

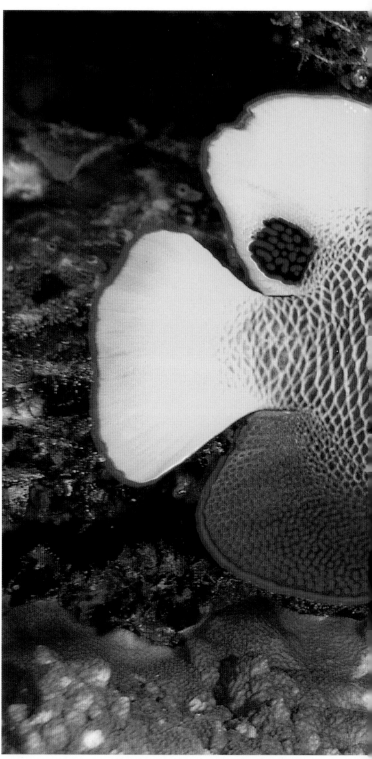

anemones and immune to the sea anemone's
paralysing venom), the *Pomacanthus* and
*Pygoplites* genera, the emperor fish; the *Odonus*
and *Balistoides* genera, the trigger fish; and
the *Scarus* genus, the parrot fish (whose
extraordinarily bright skin changes color
completely depending on age and sex, and which
feeds on the polyps and algae of the madrepores
which it bites off with its strong "beak").

*228 top Six-banded
angelfish
(Pomacanthus
sexstriatus).*

*228 bottom
Clown triggerfish
(Balistoides
conspicillum).*

*228-229
Blue-face angelfish
(Pomacanthus
xanthometopon).*

*229 bottom left
Pennant bannerfish
(Heniochus
chrysostomus).*

*229 bottom right
Racoon
butterflyfish
(Chaetodon lunula).*

Other genera commonly found on Malaysian reefs are *Acanthurus* and *Naso,* the surgeon fish (which feed on the algae that encrust the hard coral formations, using their setula-shaped teeth) and the genus *Caesio,* the so-called "fusiliers" (the small fusiform fish which live in large shoals on the outer edges of the reef). There are also many species of scorpaenids (belonging to genera such as *Pterois* or *Scorpaenopsis*), morays (particularly the *Gymnothorax* genus), and snappers of the *Lutjanus* genus; all small and medium-size predators which prefer hunting their prey among the recesses of the reef.

However, a simple list of the species more or less permanently present on the Malaysian reefs would inevitably be uninteresting and incomplete. What is fascinating is the fact that a coral reef is, in reality, a intricate mega-organism with extraordinarily dynamic internal mechanisms, essentially comprised of the constant interaction between predators and prey. At the same time it is an exceptionally balanced system, relying on fundamental elements, including sunlight, unpolluted water, appropriate water temperatures and a reliable source of nutritional materials.

230-231 *Blackspotted pufferfish* (Arothron nigropunctatus).

*231 top Map pufferfish* (Arothron mappa).

*231 center Black-saddled toby* (Canthigaster valentini).

*231 bottom Yellow boxfish* (Ostracion cubicus).

231

Unfortunately for the coral reefs, this delicate balance can be modified only too easily. While some of the outer reefs in the South China and Sulawesi Seas are as yet untouched, others are being increasingly exposed to the impact of man, through the fishing industry (industrial scale with drag nets or on a smaller scale using rudimentary explosives); through mining (the beds of the South China Sea conceal huge quantities of natural gas and oil, already being exploited near Labuan); and through tourism (for example, inappropriate resorts at Langkawi, Tioman, Redang and Pulau Sipadan).

*232 top*
*Yellow-eyed blenny (Ecsenius melarchus).*

*232 bottom Bartel's mandarinfish (Synchiropus bartelsi).*

*232-233 Goby (Eviota sp.) on a coral (Diploastrea heliopora).*

*233 top left False clown anemonefish (Amphiprion ocellaris).*

*233 top right Orange anemonefish (Amphiprion sandaracinos).*

While able to re-establish its equilibrium even after catastrophes such as a typhoon or the population boom of the crown-of-thorns starfish (*Acanthaster planci*) which feeds on coral, the reef is not, however, able to absorb the devastating environmental impacts caused by man. At present in the Malaysian waters, as in all tropical seas, man is the main source of destruction of this extraordinary and irreplaceable natural evolutive laboratory.

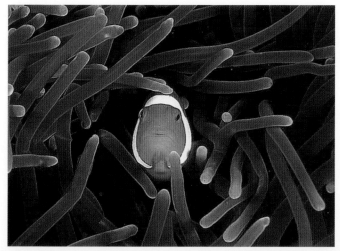

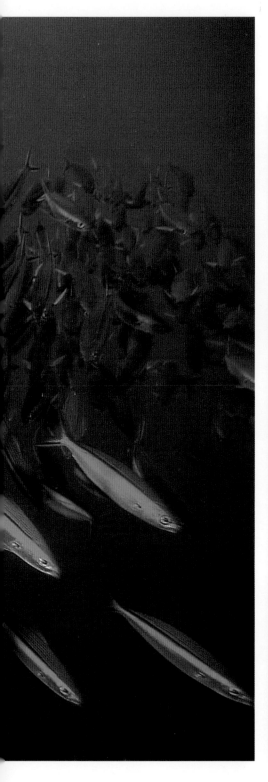

*234-235*
*Bluestreak fusilier*
(Pterocaesio tile).

*234 bottom left*
*Bumphead*
*parrotfish*
(Bolbometopon
muricatus).

*234 bottom right*
*Longfin spadefish*
(Platax teira).

*235  Boersi's*
*spadefish*
(Platax boersii).

# PREDATORS ON THE PROWL

**Lionfish**
*Pterois volitans*

**Zebra lionfish**
*Dendrochirus zebra*

**Spotfin lionfish**
*Pterois antennata*

**Stonefish**
*Synanceia verrucosa*

**Tassled scorpionfish**
*Scorpaenopsis oxycephala*

**Leaf scorpionfish**
*Taenianotus triacanthus*

**Spiny devilfish**
*Inimicus didactylus*

**Hispid frogfish**
*Antennarius hispidus*

**Freckled frogfish**
*Antennarius coccineus*

**Lunar-tailed bigeye**
*Priacanthus hamrur*

**Sargassumfish**
*Histrio histrio*

**Longnose hawkfish**
*Oxycirrhites typus*

**Coral grouper**
*Cephalopholis miniata*

**Crocodilefish**
*Cociella crocodila*

236

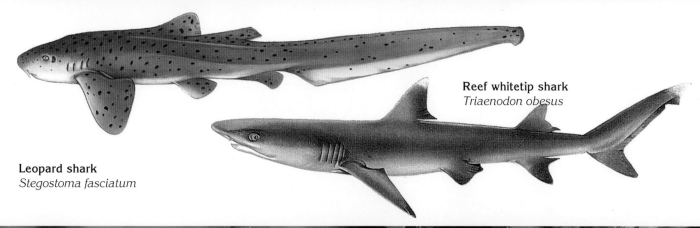

**Reef whitetip shark**
*Triaenodon obesus*

**Leopard shark**
*Stegostoma fasciatum*

*236-237 Coral grouper
(Cephalopholis miniata).*

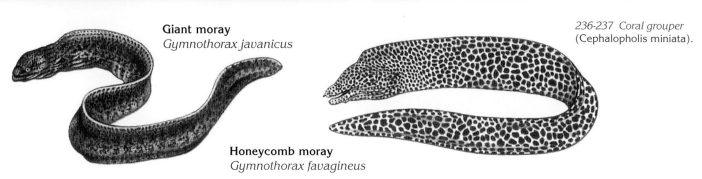

**Giant moray**
*Gymnothorax javanicus*

**Honeycomb moray**
*Gymnothorax favagineus*

Careful observation of the Malaysian sea floors - whether coral and sandy or detrital - may give the diver and marine biologist immense satisfaction. While every creature of the reef is basically a predator - from the smallest hard coral polyp to the most terrifying shark - it is indeed the most sedentary of those species devoted to hunting which offer the most interesting behaviour, the most curious skin coloring and most interesting adaptation to the environment. Relying mainly on ambush-type attacks, the most common sedentary predators of the reef leave lightning attacks and silvery garb to the inhabitants of the open sea such as sharks, tuna fish and barracudas, preferring forms of camouflage and disguise which are often able to deceive even the most experienced observers. The sedentary predators of the Malaysian reefs are often nocturnal creatures. Almost all the morays hunt outside their lairs, searching for food in the recesses of the reef during the night and even more frequently at dusk. Around dusk in the Indo Pacific, the *"Pterois"* hour occurs, when, in the unclear opaque light which precedes the dark, groups of beautiful scorpaenids leave the recesses in which they have sheltered during the day to make raids on the smaller inhabitants of the reef. They are mainly opportunists, and in general do not let good meals pass them by, even during the daylight hours.

Two of the larger predator species closely linked to the coral reefs are the leopard shark *(Stegostoma fasciatum)* and the smaller white tip reef shark *(Triaenodon obesus)*. These species are neither aggressive nor dangerous to man unless bothered. They hunt mainly at night and their diet consists chiefly of molluscs (shellfish and cuttlefish), and sleeping, medium-sized fish taken by surprise. The grey reef sharks *(Carcharhinus amblyrhynchos)*, active and audacious predators, which to all intents and purposes can be considered reef predators, are more commonly found in open water, especially between 20 and 40 meters.

*238-239 and 239 top Reef whitetip shark (Triaenodon obesus).*

*239 bottom Leopard shark (Stegostorna fasciatum).*

The numerous and often brightly-colored species of groupers, belonging to the *Epinephelus, Chromileptes* and *Cephalopolis* genera, are commonly found among the coral formations. All sea basses, even the smallest, are well known for their large appetites, satisfied by their ability to make lightning attacks with their huge jaws, and to take their prey by surprise. Slower moving, and with a really terrifying appearance, the various species of morays which live in these seas (particularly the *Gymnothorax* genus) do not deserve their bad reputation. Usually they are peaceful creatures - they generally only bite to defend themselves. They have an incredible sense of smell; this sense comes in handy during night hunts and at times leads them to investigate divers, observing them with worrying persistence. Equipped with formidable teeth, the most impressive of the morays in the South China Sea is *Gymnothorax javanicus*, almost two meters long and as thick as a man's thigh; while the most exquisitely colored is *Rhinomuraena quaesita,* with its striking shades of chromium yellow and electric blue - just a little thicker than a pencil.

241 top *Longnose hawkfish* (Oxycirrhites typus).

241 center *Ribbon eel* (Rhinomuraena quaesita).

241 bottom *White-eyed moray* (Siderea prosopeion).

240-241 *Arc-eye hawkfish* (Paracirrhites arcuatus).

However, it is among the hunters which lie in wait for their prey - the scorpionfish, stonefish and frogfish - that we find the most extraordinary species. The tropical scorpionfish belonging to the *Scorpaenopsis* genus, particularly the *oxycephala* and *venosa* species, adapt exceptionally well to the reef environment and are almost completely invisible, thanks to their camouflage coloring, but they often reveal unexpectedly bright colors under the light of the flash. Their flattened shapes, with their many appendages (dermal fringes, quills and wide-rayed fins) make them extremely difficult to see. It is in the *Pterois* genus that the camouflage function of the skin achieves its greatest elegance.

Few other creatures of the reef can boast such graceful and spectacular shapes and such perfect adaptations to the environment as *Pterois radiata*, *Pterois antennata* or *Pterois volitans*.
The smaller scorpaenids, with similar habits, such as the *Dendrochirus zebra*, *biocellatus* and *brachypterus* are also extraordinarily elegant and have an exceptional ability to mimic their environments.

One certainly does not think of elegance when observing another scorpaenid, the *Synanceia verrucosa* stonefish, whose shape and color provide a superb camouflage on the seabed. Its sedentary habits allow colonies of algae and hydroids to attach to its body.
This is one of the few types of fish which is genuinely dangerous to divers because of its exceptional ability to mimic its surroundings, and the powerful poison it injects into its victim through the hollow spines of its dorsal fins. The *Scorpaenopsis diabolus* is very similar and almost as dangerous. Harmless to man, the small and incredibly well-camouflaged leaf scorpionfish *(Taenianotus triacanthus)* found in the Malaysian seas, whose skin varies from white to brown through bright yellow, green and purple, and the frogfish *(Antennarius moluccensis)*, closely related to the Mediterranean devil-fish, are almost invisible among the corals

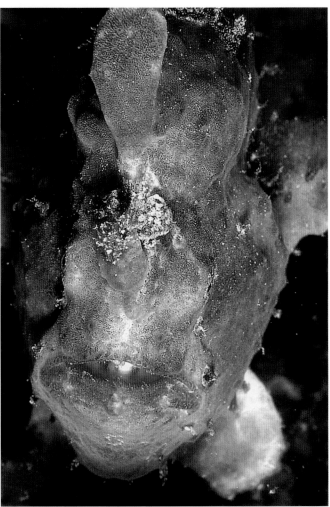

and candelabra sponges where they are often seen resting during the day. Like the leaf scorpionfish, the Mediterranean devil-fish lives in a small, well-defined area which it does not leave over the years and the colors of its skin tend to mimic its surroundings.

**Hawksbill turtle**
*Eretmochelys imbricata*

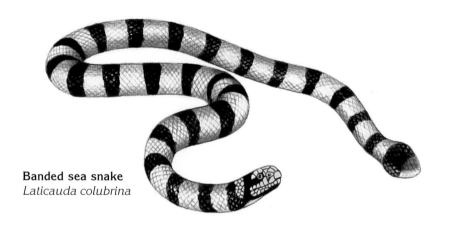

**Banded sea snake**
*Laticauda colubrina*

**Green turtle**
*Chelonia mydas*

The coasts of peninsular Malaysia and Borneo are still mostly uninhabited and largely unexplored. The various coastal ecosystems include the beaches of sand and coral fragments (the main system); basaltic cliffs of ancient volcanic origin; the mudflats or expanses of silt or mud periodically uncovered by the falling tide; and, above all, the huge mangrove forests in association with this habitat and close to the estuaries, which are characteristic of coastal regions. Many of these environments, which are all so different, offer excellent opportunities for feeding, reproduction and rearing of young of various types of reptiles: turtles, primarily, but also crocodiles and snakes. The giant leatheback turtle *(Dermochelys coriacea)*,

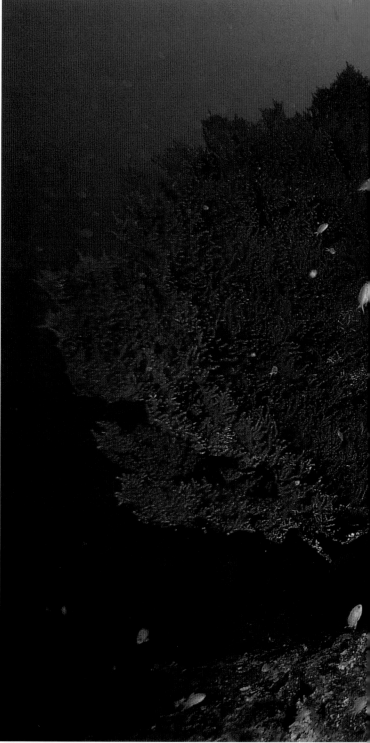

the green turtle *(Chelonia mydas)*, and the hawksbill turtle *(Eretmochelys imbricata)* are (or were) common in Malaysian waters. The leatherback turtle is of strictly pelagic habits; it eats only jellyfish despite its mass and can grow to more than 700 kilos in weight. It was once very common during the mating season and laid its eggs on the sandy beaches of the peninsular state of Terengganu, but its numbers are now unfortunately much reduced throughout the world and scientists now fear for the survival of the species. This majestic giant, with its antediluvian appearance, is present throughout the world, but it is also the victim of the ever-present plastic bags floating on

the surface (which they eat mistaking them for their usual food and suffocate), of the fishing nets placed by the fleets of fishing boats, of pollution of all types near their mating and egg-laying sites and, finally, of ruthless hunting. The other two species are much more common at various, at least theoretically, protected areas and are much easier to come across during dives. The green turtle *(Chelonia mydas)* prefers to feed on the beds of sea grass near the coast but it is more common to see it near isolated offshore locations with sandy beaches (such as Pulau Sipadan) where it hunts for a partner or makes its way during the night towards land to dig its

248-249
*Hawksbill turtle*
(Eretmochelys
imbricata).

large nest in the sand and lay its eggs. It is a
shy, docile animal which swims elegantly and
peacefully, and the older of the species grow fairly
large. It is not uncommon to encounter turtles
which have somehow survived an encounter with
tiger sharks *(Galeocerdo cuvieri)* with part of a
limb or even a part of the shell cleanly bitten off
by a predator's jaws. The term green, usually
used to describe the species, does not refer to
the color of the animal but to its fat - the
main ingredient of the deplorable turtle soup
too often requested by western tourists looking
for unusual, new gastronomic experiences.
The *Chelonia mydas* is still ruthlessly hunted

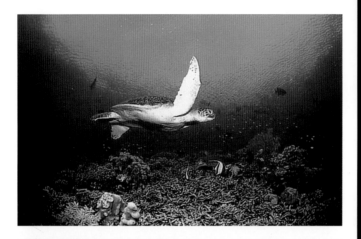

*250-251*
*Green turtle*
(Chelonia mydas).

throughout Indonesia and in most of Southeast Asia, especially near Bali. Another reason for the gradual disappearance of the species is the indiscriminate collection of its eggs, which are eaten in huge numbers by locals. Also in danger, because of the beauty of its translucent shell, is the small hawksbill turtle *(Eretmochelys imbricata)* more common to the reef environment than the other species. Its favourite foods are sponges and soft corals which technically makes it a carnivore. This species can be easily approached during dives, but since it is a protected species it should not be interferred with. Third-rate divers have been known to disturb turtles, and worse still, grab hold of them. Despite its huge muscle power and its ability to stay underwater for long periods, the

turtle has little resistance to stress and often suffers a heart attack and drowns just after it is the victim of these foolish pranks.

For obvious reasons, it is highly inadvisable to disturb or try to handle other reptiles present in the Malaysian archipelago such as sea snakes (very common in some areas of the Indo Pacific) and saltwater crocodiles (which now live only in more remote and wilder locations). The most common sea snake encountered in Malaysian waters is the striped krait *(Laticauda colubrina),* which belongs to the *Elapidae* family, and is in fact a close relation of the land cobra, having a strong neurotoxic venom in common. This animal, growing to more than one and a half meters long, has exceptionally elegant grey-blue

and black stripes and its face is bordered with pale yellow. Its favourite foods are the fish and crabs which live in vast numbers on the mudflats and are easily found near the coast. It is a very curious species, by no means shy, and it often approaches divers to study them close up. During the mating season in spring, it is also fairly easy to see them after sunset when they come out of the water and search the beach for a partner. Much more dangerous, but now fairly rare, is the saltwater or estuarine crocodile *(Crocodylus porosus),* which grows to an enormous length of up to 7 meters, and can be extremely aggressive (it represents one of the main risks to swimmers in many areas in northern Australia). It lives in salt water and the estuaries of large rivers, travelling along them for kilometers, as well as in the mangrove forests of almost all Southeast Asia. This creature has also been seen in the open sea, several kilometers from the coast and on the coastal reefs. It is one of the few predators which, in some circumstances, could represent a mortal danger for divers.

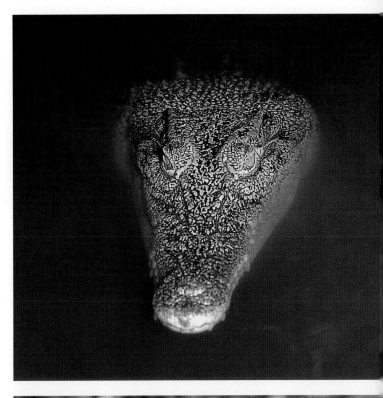

*252 and 253 bottom Banded sea snake* (Laticauda colubrina).

*253 top Saltwater crocodile* (Crocodylus porosus).

# KNIGHTS IN ARMOUR

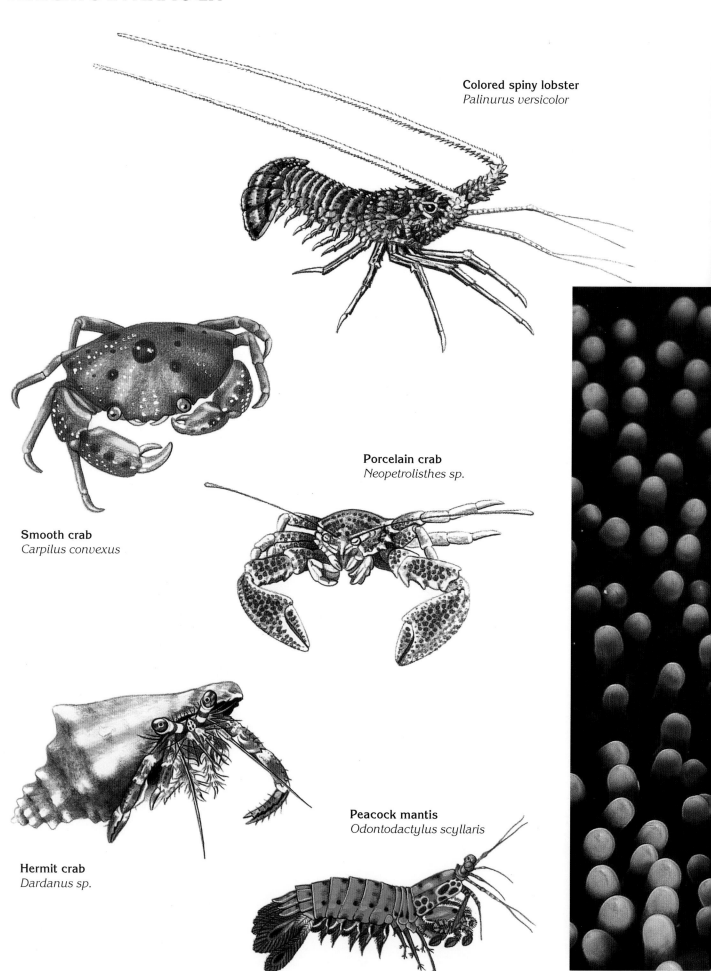

**Colored spiny lobster**
*Palinurus versicolor*

**Porcelain crab**
*Neopetrolisthes sp.*

**Smooth crab**
*Carpilus convexus*

**Hermit crab**
*Dardanus sp.*

**Peacock mantis**
*Odontodactylus scyllaris*

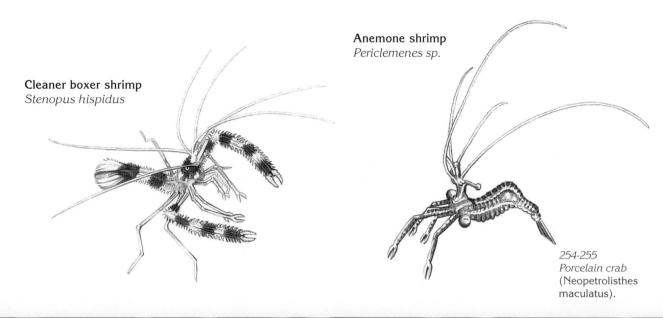

**Cleaner boxer shrimp**
*Stenopus hispidus*

**Anemone shrimp**
*Periclemenes sp.*

*254-255*
*Porcelain crab*
(Neopetrolisthes
maculatus).

The Malaysian coral reefs give shelter to a vast number of species of exceptional interest to divers, biologists and photographers, always on the lookout for new and surprising subjects to study and record. Shellfish, some of the most difficult inhabitants of the reef to observe, are often also the most colorful and interesting. Shy, mainly nocturnal creatures, they are unfortunately among the most hunted by predators. They are the favourite food of many molluscs (such as octopus, cuttlefish and squid) and of many fish (especially scorpaenids and sea basses) and man also enjoys their flesh. They are fairly small in size. Crabs, shrimps and lobsters move among corals and alcyonarians like little knights in medieval armour. Like the knights, they are protected by a strong outer shell, and, like them, they are often doomed to death if surprised in the open. During the daytime, shellfish which can be encountered without particular problems (since they are all are fairly small in size) are the cleaner shrimps belonging to the *Stenopus* genus. The *hispidus* species, more common than the others, is easily recognized by its long antennae and the elegant white and red striped casing, often awaiting "clients" in special cleaning stations, frequently visited by reef fish that recognize them by the rhythmic swaying of their long antennae.

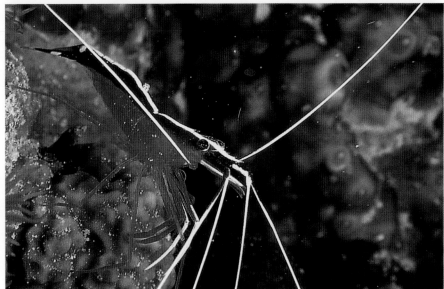

256-257
*Dancing shrimp*
(Rhynchocinetes
sp.).

*257 top  Cleaner
boxer shrimp*
(Stenopus
hispidus).

*257 center  Cleaner
shrimp* (Lysimata
amboinensis).

*257 bottom
Anemone shrimp*
(Periclemenes sp.).

With a more markedly nocturnal nature, the shrimps belonging to the *Rhynchocinetes* genus are the most easily identified. Their large eyes reflect the light of the divers' torches as they peep out from the cracks and recesses in which they often hide. Those belonging to the Saron genus are rather difficult to find, but are pursued by photographers because of their particular profile (many have a curious setula mane on their back), and their brightly-colored bodies with their curious geometric designs. Easy to observe during the day, are the symbiont shrimps (mainly belonging to the *Periclemenes* genus) whose favourite habitats include the basal peduncle (among the fiercely venomous tentacles of the huge Actinia), and the pinnulated arms of the crinoids. The cleaner shrimp generally stands out with their extraordinarily

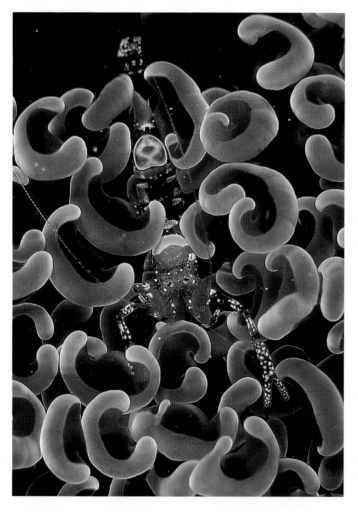

colorful skin which helps their potential clients to recognize them (also these shrimps tend to clean fish of parasites and particles of mucus and dead skin). The symbiont shrimps (which often share the crinoid on which they live with other shellfish of the *Allogalathea elegans* species, and with small fish and serpentine starfish) have an extraordinary camouflage livery which allows them (by a mechanism still to be understood) to permanently take and perfectly mimic the colors of their host and become almost invisible not only to their potential predators, but also to most divers.

*258  Anemone shrimp*
(Periclemenes sp.).

*258-259*
*Anemone shrimp*
(Periclemenes sp.)
*in bubble coral*
(Plerogyra flexuosa).

*259 bottom*
*Harlequin shrimp*
(Hymenocera sp.).

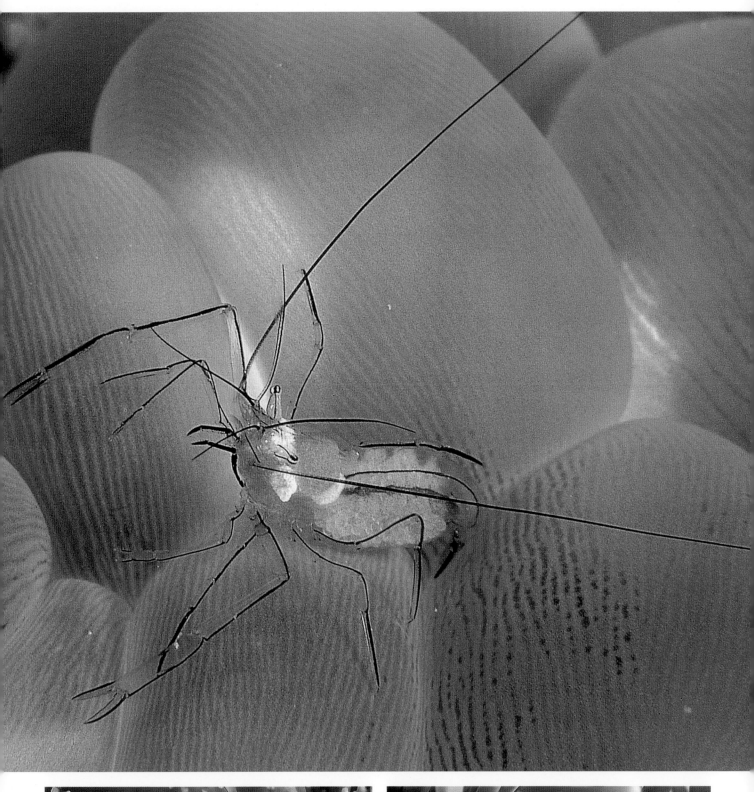

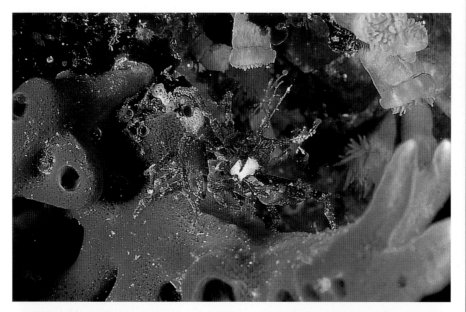

Among the tentacles of the anemones it is often easy to observe some isolated examples or pairs of elegant porcelain crabs of the *Neopetrolisthes* genus; these are also symbiont shellfish which exploit the venomous capacity of their host to protect themselves against predators. They feed on waste products produced by the anemones and plankton. Other small crabs (generally a few centimeters in diameter) belonging to the *Trapezia*, *Vymo* and *Teralia* genera, prefer to take refuge among the branches of the hard coral formations - it is not unusual to see them hiding among the thick colonies of *Pocillopora*. The *Thalamita* genus is commonly found on muddy sea beds and in salt water. It is more difficult to find other species of larger shrimps or crabs, with the exception of various species belonging to the *Carpilius* genus, which are fairly common on the reef during the night.

*260 and 261 bottom Decorator crabs.*

*260-261 Crab of the* Majidae *family.*

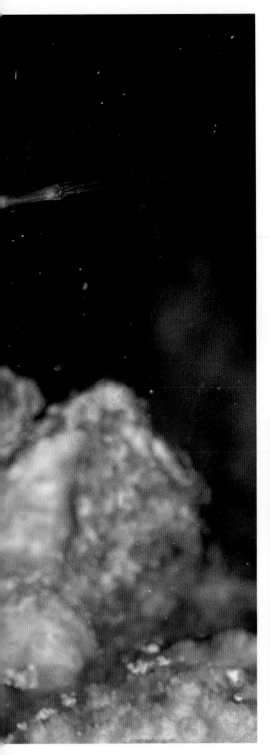

The most common of the various species of lobster is the *Palinurus versicolor,* with its splendid shiny, turquoise, light blue and bright pink shell. In places where the populations have not been damaged by intensive fishing, it is fairly common to see their long white antennae sticking out of recesses on the reef during the day. This creature is another opportunist with nocturnal habits, which feeds on live or dead fish, molluscs and shellfish. Finally, the peacock mantis *(Odontodactylus scyllaris),* with its predatory habits and beautiful carapace, is a very active crustacean with daytime habits. Unlike its conspecifics, it often abandons its hole (a vertical tunnel dug into the detrital substratum) to hunt its prey.

*262-263*
*Peacock mantis*
(Odontodactylus scyllaris).

*262 bottom left*
*Colored spiny lobster* (Palinurus versicolor).

*262 bottom right*
*Tropical mantis*
(Odontodactylus sp.).

*263 Hermit crab*
(Dardanus sp.).

The tropical squills are extremely sophisticated predators whose "grabbing" front limbs - similar to those of a praying mantis - are able to extend and then close again in a fraction of a second, using the principle of the flick-knife, to impale the unfortunate prey on the serrated and aculeate edges of its "forearms". The coral reefs of the Malaysian seabed offer shelter to various species belonging to the *Odontodactylus* genera ("fanged fingers"), some of fairly large sizes (up to 40 centimeters long), however, the *scyllaris* species is certainly the most spectacularly colorful and is also the most active and easiest to observe.

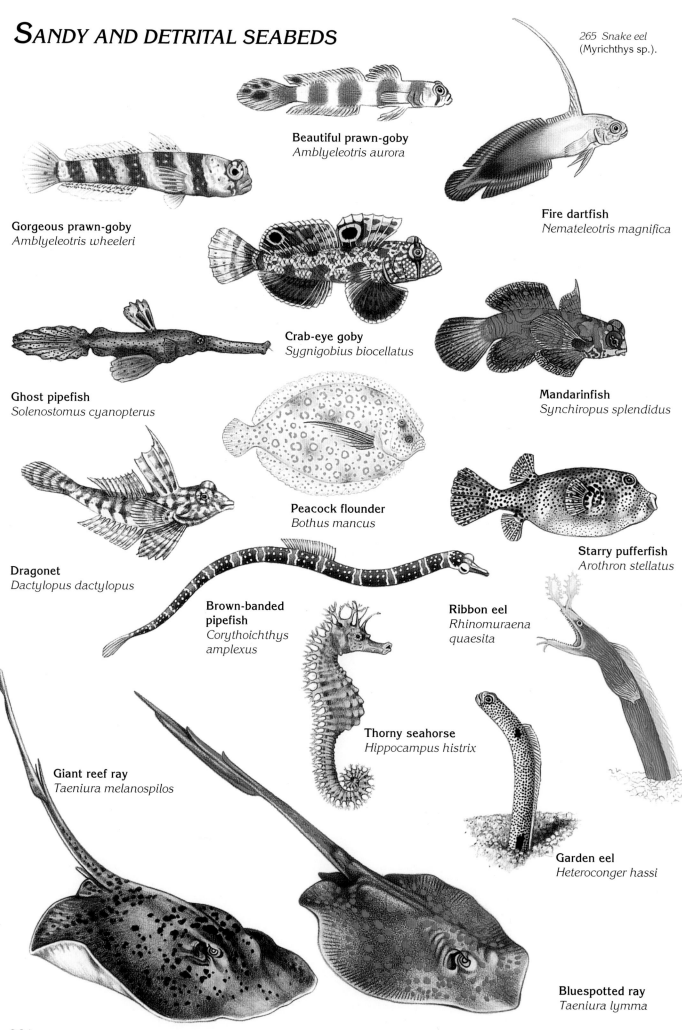

# SANDY AND DETRITAL SEABEDS

265 Snake eel
(Myrichthys sp.).

**Beautiful prawn-goby**
*Amblyeleotris aurora*

**Gorgeous prawn-goby**
*Amblyeleotris wheeleri*

**Fire dartfish**
*Nemateleotris magnifica*

**Crab-eye goby**
*Sygnigobius biocellatus*

**Ghost pipefish**
*Solenostomus cyanopterus*

**Mandarinfish**
*Synchiropus splendidus*

**Peacock flounder**
*Bothus mancus*

**Dragonet**
*Dactylopus dactylopus*

**Starry pufferfish**
*Arothron stellatus*

**Brown-banded
pipefish**
*Corythoichthys
amplexus*

**Ribbon eel**
*Rhinomuraena
quaesita*

**Thorny seahorse**
*Hippocampus histrix*

**Giant reef ray**
*Taeniura melanospilos*

**Garden eel**
*Heteroconger hassi*

**Bluespotted ray**
*Taeniura lymma*

However inhospitable the sandy, muddy or detrital beds may seem at the outset to inexperienced divers, they represent the ideal habitat for an extraordinary variety of benthic species characteristic of the Malaysian coasts. The first creatures the more careful observer will notice are a wide variety of small multi-colored gobies belonging to the *Cryptocentus* genus as they peep out of their holes dug in the sand, or hang about nearby. These are tiny, fascinating fishes with marked territorial habits; each indicates its zone of influence, at times just a few square decimeters, with a great fanning out and stretching of tiny colorful fins. Extremely attentive to what happens around them, they are always ready to dive like lightning into their holes if they suspect that there is a danger, and it takes great patience and care to get close enough to be able to observe them easily. As they watch these creatures, the more attentive divers will soon note the fascinating interaction between the gobies and the shrimps of the *Alpheus* genus - like small bulldozers, intent on obsessively cleaning the inside and the opening of the tunnel dug in a layer of debris, which builds up continually. Generally characterized by a striped body, the symbiont shrimp is almost completely blind and communicates with its host, the goby - which on the contrary has excellent sight - with light touches of its antennae. The small fish on the other hand communicates impending danger to the shrimp with rapid and light touches of its tail. The dragonets of the *Dactylopus* genus are also easy to find on sandy beds; these creatures have a spectacular, multi-colored dorsal fin used to communicate with their conspecifics. Other fairly common inhabitants of these stretches of sand are the garden eels of the *Heteroconger* genus, small eels whose front third usually sticks out of the sand like a small periscope projecting into the current. The garden eels often gather together in large colonies of several hundred or more, which rhythmically sway in the current like little stalks of grass - hence their name. They are very shy animals and difficult to find. Often characterized by a spotted or striped skin, they feed on the plankton carried past by the current and do not often abandon the safety of their holes.

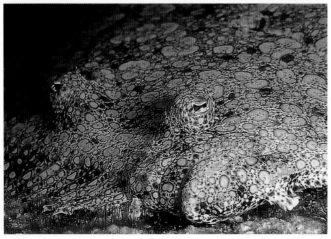

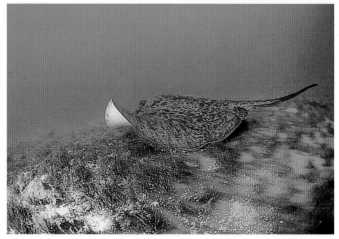

The so-called sand eels or snake eels of the *Myrichthys* genus are much larger and can reach 80-100 centimeters in length. They are nocturnal predators on the sandy and detrital beds of the reef, which is the same ecological niche occupied by the morays. By day it is sometimes possible to observe their pointed faces, with their characteristic tubular nostrils, sticking out above the substratum. Larger rays and stingrays (from around 30 centimeters to more than 2 meters wide) are among the most spectacular inhabitants of the sandy beds and belong mostly to the *Dasyatis* and *Taeniura* genera. These rays are nocturnal predators (they feed on shell fish and

*268-269 Dragonet (Dactylopus dactylopus).*

*268 bottom left Peacock flounder (Bothus mancus).*

*268 bottom right Giant reef ray (Taeniura melanospilos).*

*269 Bluespotted ray (Taeniura lymma).*

small fish) which spend most of the day half buried under a light layer of sand, relying on their flattened profile and on their camouflage to hide them from other predators (hammerhead sharks seem to have a real passion for these rays). Almost all these species have a serrated barb near the base of the tail, connected to a venom gland, which contains a toxic secretion potentially dangerous to man as well. The flat form typical of these rays is also shared by species of sole, and spotted turbots such as the *Bothus pantherinus*.

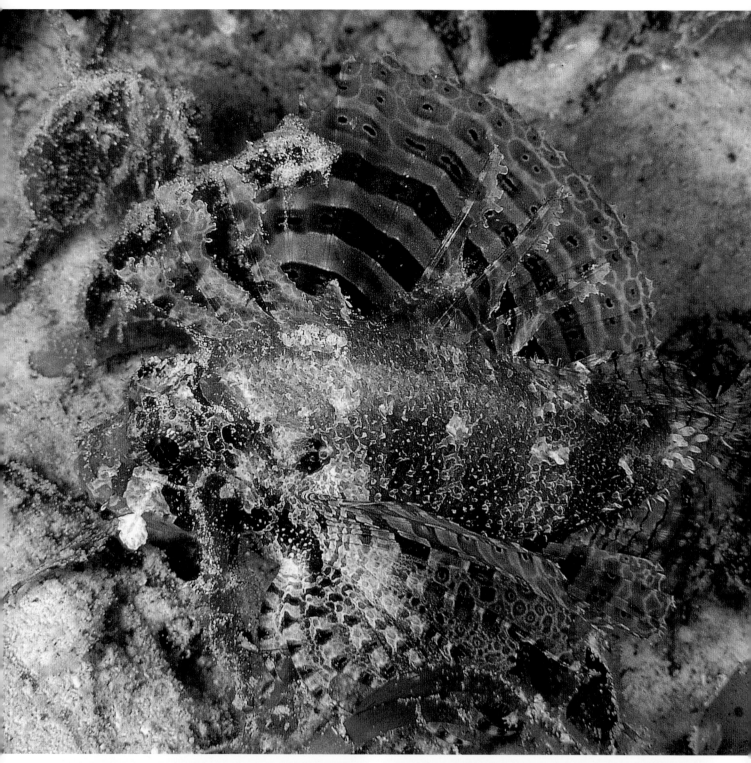

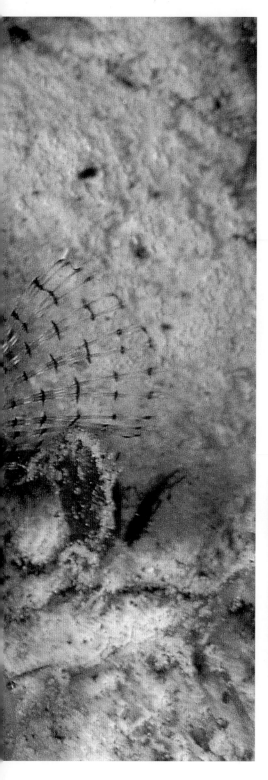

More characteristic of sandy beds, especially those areas rich in algae and sponges, are the numerous frogfish, pipefish of the *Corythoichthys* genus, and some species of seahorses of the *Hippocampus* genus. However these creatures are not often seen because of their extraordinary ability, acquired over millions of years of adaptation, to mimic their surroundings, and because the environments they prefer are not usually visited by amateur divers. Even though they also have an exceptional ability to mimic their surroundings, the various genera of scorpaenids are much easier to find since they live on the barrier reef and detrital beds where the reef has been partially destroyed by storms or man. The stonefish *(Synanceia verrucosa)* has astonishing camouflage and can be seen at times in water just a few centimeters deep. It and the numerous members of the *Scorpaenopsis* genus, which includes very different species to those which prefer

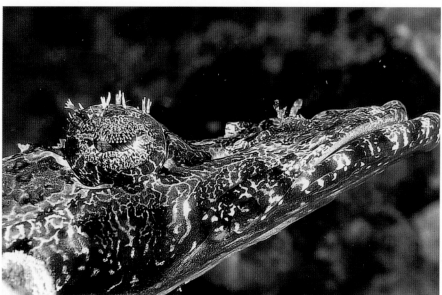

the coral reef environment, are fascinating bottom dwellers. Another sedentary predator which prefers to hide on sandy or muddy beds is the devil scorpion fish *(Inimicus didactylus),* with its brightly-colored pectoral fins. Characteristic of detrital beds, the large crocodilefish *(Cymbacephalus beauforti)* has an impressive arabesque skin that makes it almost invisible when it lies immobile on the seabed waiting for its prey. It is not a scorpaenid but has camouflaging colors and an ambush hunting system similar to the scorpionfish.

Another typical predator of the detrital beds, characterized by its small size and its splendidly-colored skin, is the beautiful ribbon eel *(Rhinomuraena quaesita)*, which sometimes can be seen as it peeps out of its hole. On the most decayed beds of the Indo Pacific live many other smaller, but no less colorful, species of ribbon eel. Among the soft and hard corals of the reef, where there is plenty of sunlight and fairly shallow water, is the favourite habitat of the marvellous ghost pipe-fish of the *Solenostomus* genus, and the colorful mandarinfish of the *Synchiropus* genus.

*272 top left and 273 Ghost pipefish (Solenostomus cyanopterus).*

*272 bottom left and right Harlequin ghost pipefish (Solenostomus paradoxus).*

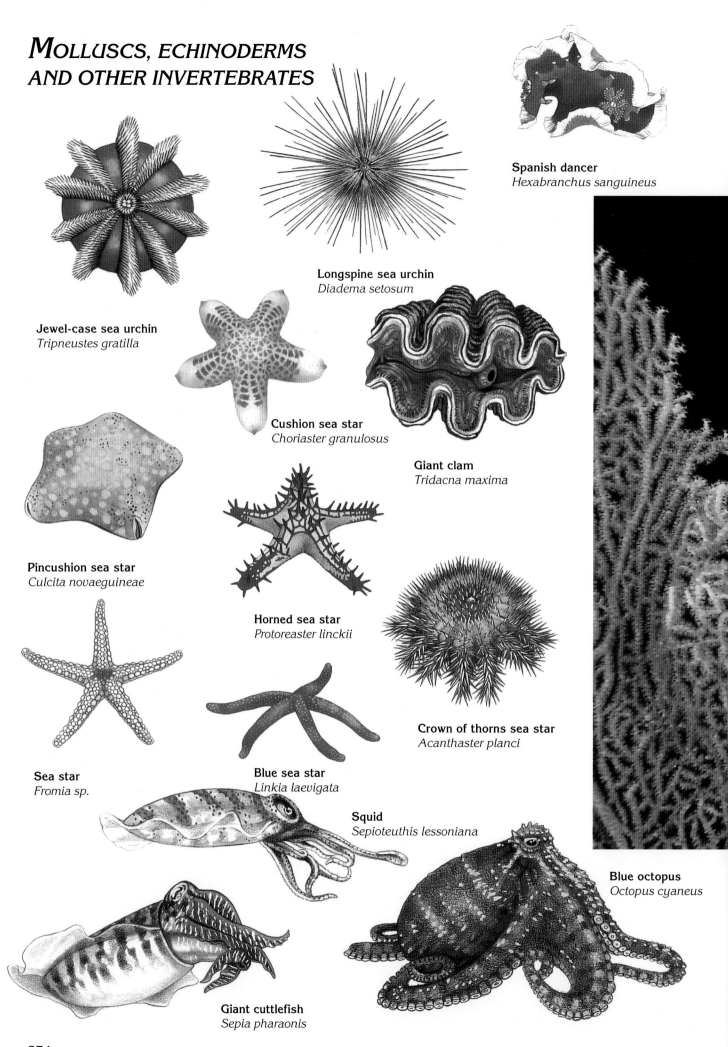

# MOLLUSCS, ECHINODERMS AND OTHER INVERTEBRATES

**Spanish dancer**
*Hexabranchus sanguineus*

**Longspine sea urchin**
*Diadema setosum*

**Jewel-case sea urchin**
*Tripneustes gratilla*

**Cushion sea star**
*Choriaster granulosus*

**Giant clam**
*Tridacna maxima*

**Pincushion sea star**
*Culcita novaeguineae*

**Horned sea star**
*Protoreaster linckii*

**Crown of thorns sea star**
*Acanthaster planci*

**Sea star**
*Fromia sp.*

**Blue sea star**
*Linkia laevigata*

**Squid**
*Sepioteuthis lessoniana*

**Blue octopus**
*Octopus cyaneus*

**Giant cuttlefish**
*Sepia pharaonis*

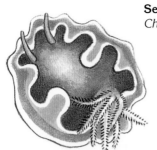

**Sea slug**
*Chromodoris sp.*

**Sea slug**
*Chromodoris bullocki*

**Four banded sea slug**
*Chromodoris quadricolor*

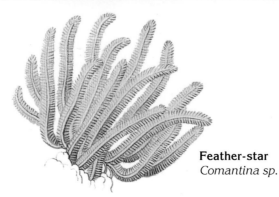

**Feather-star**
*Comantina sp.*

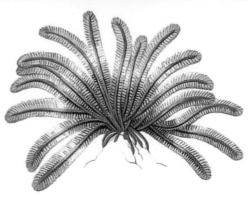

274-275  *Feather star*
(Comantina sp.).

275

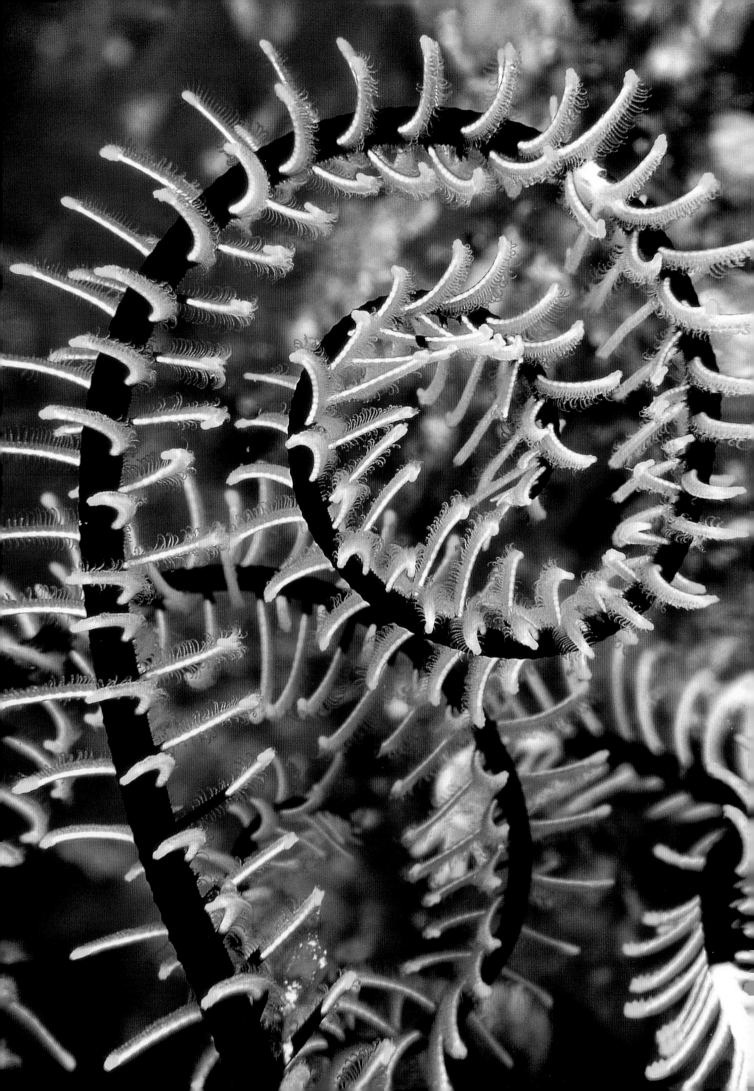

The reefs of the Malaysian Indo Pacific provide the ideal environment for an extraordinary variety of invertebrates. These are mostly primitive creatures, relatively unsophisticated from an evolutionary point-of-view, with the exception of the octopus, cuttlefish and squid. Usually these intertebrates are very small, with colors and shapes that are often among the most spectacular of the tropical underwater world. Indeed, the barrier reefs themselves have been formed by an infinite number of tiny coral polyps and are occupied by an extraordinary variety of sessile organisms such as sponges, sea fans and alcyonarians (soft corals). In the saraband of life forms passing before the eyes of the diver on a Malaysian reef are the outstandingly wonderful patterns and colors of primitive organisms such as nudibranchs (sea slugs), starfish or crinoids (feather stars). These creatures, easy to identify and photograph, have shapes and colors that at times reach pure abstraction. Color and shape play precise roles in the animal and plant kingdoms, demonstrating an adaptation to the environment which has developed over millions of years. The brightest colors of many invertebrates act as a warning, indicating to their potential aggressors that their prey is poisonous. Many species of nudibranchs of the *Phyllidia* or *Chromodoris* genera absorb and store in their bodies the venomous properties of their prey. The long fragile spines of the *Diadema* sea urchins have a defensive or perhaps even an offensive role although the frogfish, which feeds on them, has discovered how to avoid the spines of its favourite food. Taking the end of a spine gently in its mouth, the fish turns the sea-urchin over and swallows the sea urchin, starting from the relatively unprotected underside. The feather stars (ancient organisms with nocturnal habits, which filter sea water with the pinnules edging their "arms") do not appear to have natural enemies and often shelter an entire small community of shellfish and symbiont fish among their fragile arms. Commensal relationships reach their peak among the huge sea anemones of the *Heteractis* genus, and the various species of clown anemone fish of the *Amphiprion* genus which live among their tentacles. The center of their distribution is the central Indo Pacific region.

278 left and 279
Giant clam (Tridacna
maxima).

278 right Crown
of thorns sea star
(Acanthaster planci).

During the night on the Malaysian reef it is
not uncommon to encounter many other
spectacular invertebrates of biological interest.
Bivalves with multi-colored mantles encrust the
walls (especially *Lopha cristagalli* and *Spondylus
aurantius*), often attaching themselves to
organisms such as sea fans, while other
fascinating bivalves such as the large *Tridacna
maxima* settle on the hard coral formations.
As well, colorful sea stars move among these
same formations - the beautiful *Linckia laevigata*,
usually a bright blue color, the cushion
sea star *(Culcita novaeguineae)* with its
characteristic rounded form, and the smaller,

elegant *Fromia monilis* with its splendid colors -
are unmistakable. Among the sea urchins which
are most easily seen in the recesses of the coral
reefs are the spectacular *Diadema*, with its
distinctive long fragile spines and the beautiful
*Tripneustes gratilla* sea urchin with its blue or
green velvety "segments" and its short yellowish
spines. Much more dangerous, and potentially
lethal for man, are the sea stars belonging to the
*Toxopneustes* genus which are often covered
by algal debris and fortunately are only found
on muddy beds.

280  Sea star.

281 top left
Blue sea star
(Linckia laevigata).

281 bottom left
Sea star
(Linckia multiflora).

281 right
Cushion sea star
(Choriaster
granulosus).

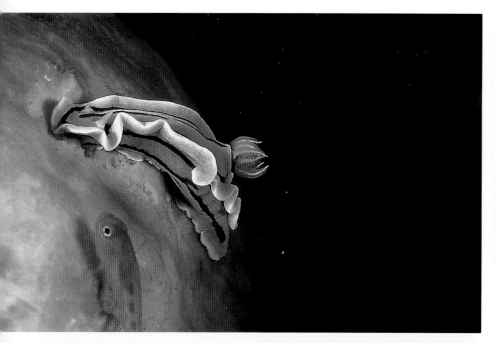

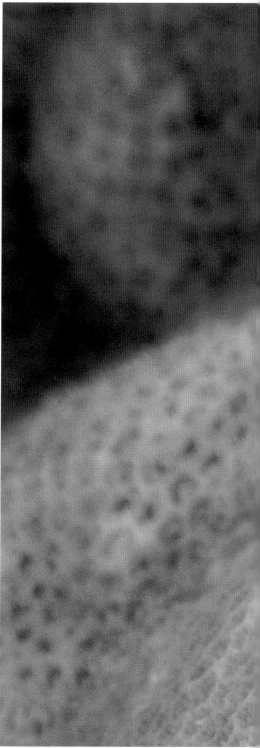

*282 top  Sea slug* (Chromodoris willani).

*282 bottom Flatworm* (Pseudoceros affinis).

*282-283  Sea slug* (Chelidonura amoena).

*283 bottom left Sea slug* (Phylidia varicosa).

*283 bottom right Sea slug* (Chromodoris bullocki).

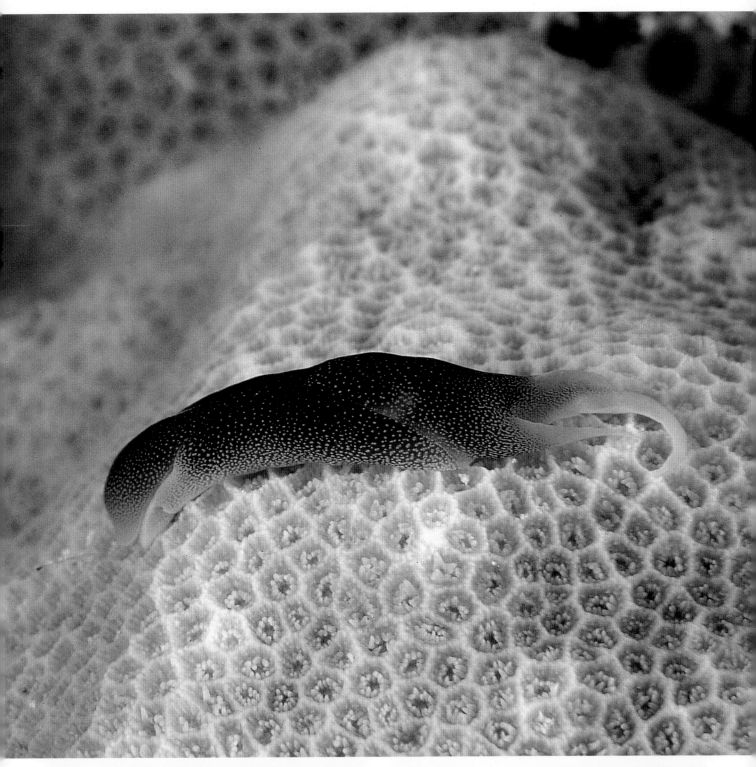

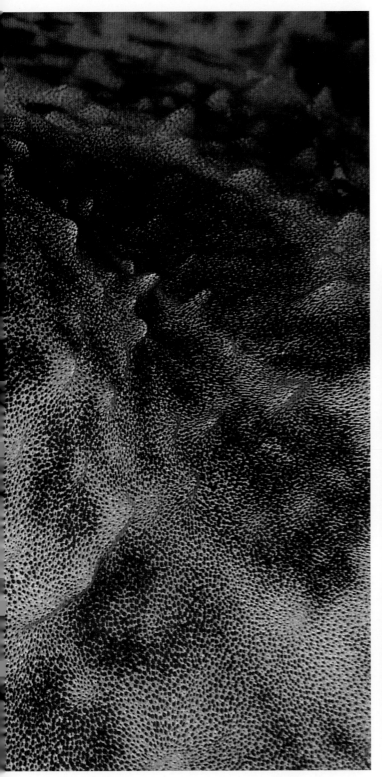

The intelligent cephalopod molluscs - octopus, cuttlefish and squid - will particularly fascinate the diver if he happens to come across them. These creatures are hunted by sharks, sea basses and morays more in tropical waters than in temperate waters. These molluscs, which prefer coastal waters with sandy or muddy beds, are fascinating creatures with a remarkable ability to mimic their surroundings. The octopus and cuttlefish are usually found in the benthic environment, and reach remarkable dimensions - the giant cuttlefish *(Sepia pharaonis)* can grow to more than 1 meter in length, excluding the tentacles. They can express their feelings very clearly and communicate with individuals of the

*284-285, 284 bottom right and 285 Giant cuttlefish (Sepia pharaonis).*

*284 bottom left Blue octopus (Octopus cyaneus).*

same species through sensory signals (octopus and cuttlefish change the texture and color of the skin, while squid use bioluminescent organs to generate intermittent luminous signals). Species which can be found in coastal waters, generally during the night, include the giant cuttlefish *(Sepia pharaonis)* mentioned above, and *Sepioteuthis lessoniana,* which has a gregarious nature and lives in shallow coastal waters. Great quantities of the latter are caught by local fishermen.

# THE ASSOCIATIONS

It is surprising to note how many commensal associations occur between different species within the realm of the coral reef - where the main activity consists of eating your neighbours. There are many interesting examples of these relationships. The participants are generally easy to find and the more observant divers will often see some fascinating interactions in Malaysian

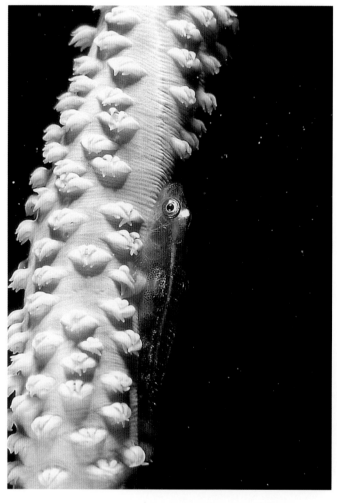

waters. In open water there is often an association between pilot fish and the large pelagic sharks (such as the *Carcharhinus longimanus),* and the well-known shark sucker *(Naucrates ductor)* and the sea turtles. The most common and easily observable example of living together (commensalism) is the curious association between the clown anemone fish of the *Amphiprion* genus (many species are represented in these waters) and the large sea anemone of the *Heteractis, Stoichachtis* and *Radianthus* genera.

*286 left Sea whips* (Juncella sp.) *and goby* (Bryaninops sp.).

*286-287 Barrel sponge* (Xestospongia testudinaria) *and feather star.*

*287 bottom left Reef whitetip shark* (Triaenodon obesus) *and sharksucker* (Echeneis naucrates).

*287 bottom right Green turtle* (Chelonia mydas) *and sharksucker* (Echeneis naucrates).

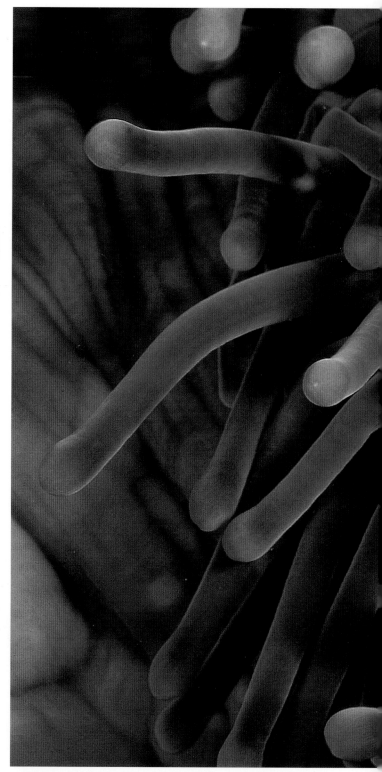

Having gradually become immune to the sting of the nematocysts of the anemone, whose paralysing power is neutralized by the mucus with which the fish is covered, the small *Amphiprion* never moves too far away from its protective host. It is not uncommon to observe large sea anemones, each with five or six clown anemone fish. These lay their eggs on the basal disk of the coelenterate (the eggs are thus well protected against ill-intentioned predators).

Anemone fish do not hesitate to courageously attack divers who come too close, painfully biting what must appear to them to be really gigantic creatures. The welcoming embrace of the large sea anemone's tentacles are also used to protect the young fish: it is not uncommon to see an entire family of *Amphiprion* - male, female and young - living among the tentacles of a single sea anemone, where it appears to have exclusive rights.

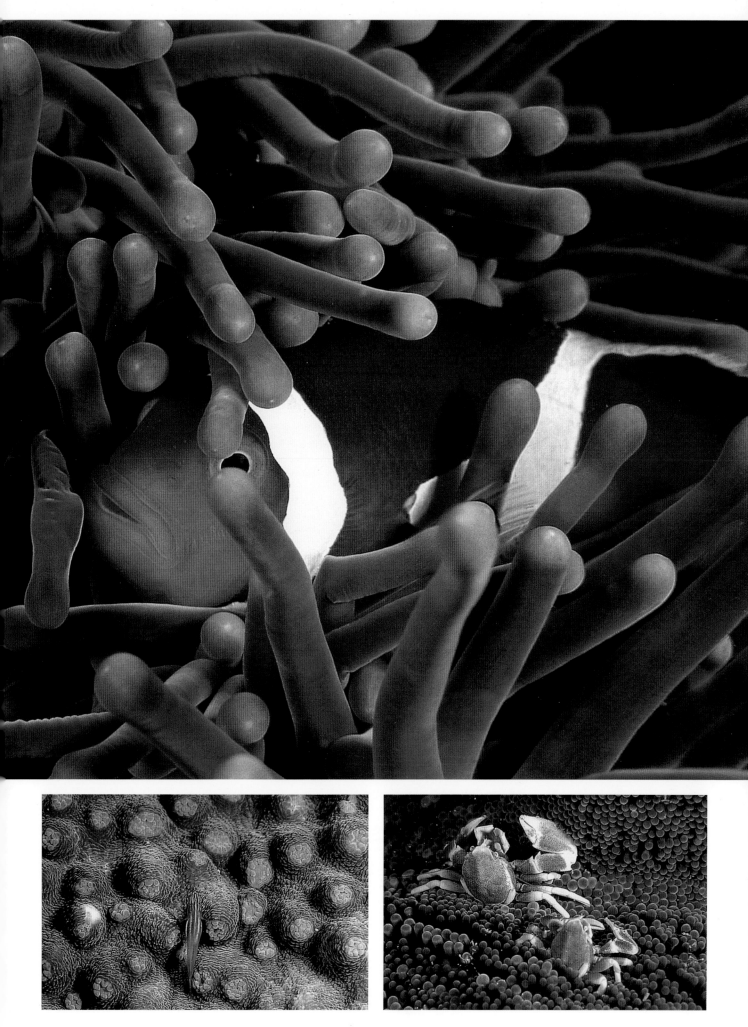

Another example of living together (this time the association apparently benefits the guest) is the community of small animals which is found among the pinnulated arms of the feather star. By moving the base very carefully, and opening the feather star up a little, it is possible to see the tiny squat lobster *(Allogalathea elegans)* with its distinctively-beaked face and its long thin nippers. Small shrimps of the *Periclemenes* genus may also be seen, which spend their life sheltered among the arms of the feather stars and mimic their colors in an astonishing and impressive way, even reproducing the alternating colors of their hosts. The small shellfish are not the only guests of the crinoid; they often share their refuge with small fish (which also have an exquisite camouflage livery),

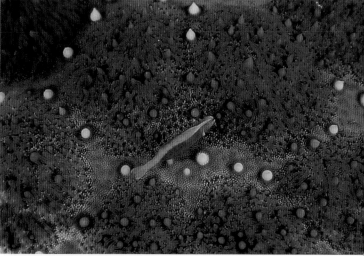

or with the small serpentine sea star of the *Ofiura* genus. The *Diaemichthys lineatus* fish has similar habits and chooses to live only among the long and dangerous spines of the *Diadema* sea urchins. Another example of commensalism - but generally in deeper water - is shown by the tiny, semi-transparent goby, belonging to the *Bryaninops* genus, which spends its entire life along the flexible whip-like coral colonies of the *Elisella* and *Juncella* genera. There are many small shellfish and cowries which - duly camouflaged - choose to reside in the delicate colonies of alcyonarians belonging to the *Dendronephtya* genus.

In these two cases the little guest is welcome as long as it keeps the organism offering it shelter clean, by feeding on mucus, parasites and waste products. However, this notion of living together is perhaps a little exaggerated.

A more active type of commensalism and one of greater mutual benefit is displayed by the gobies of the seabed and their guests, the blind shrimps of the *Alpheus* genus. Active, with excellent sight and always alert, the goby uses a tunnel dug into the sand as the center of its territory, where it quickly hides when a real or presumed danger approaches. However, the sand is easily dislodged and this is where the shrimp (or sometimes a couple of shrimps) becomes useful. It spends most of its life cleaning the tunnel and reinforcing the

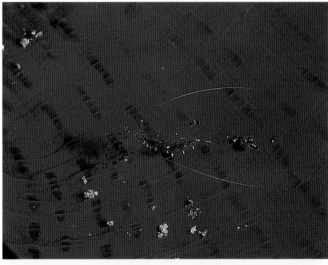

structure. There are mutual advantages: while the goby can rely on a clean and protected tunnel, the shrimp takes advantage of the alertness of the fish, which warns the shrimp of impending danger with a careful flick of the tail fin before diving into the hole. More observant divers will note that the physical contact between the two creatures is continuous.

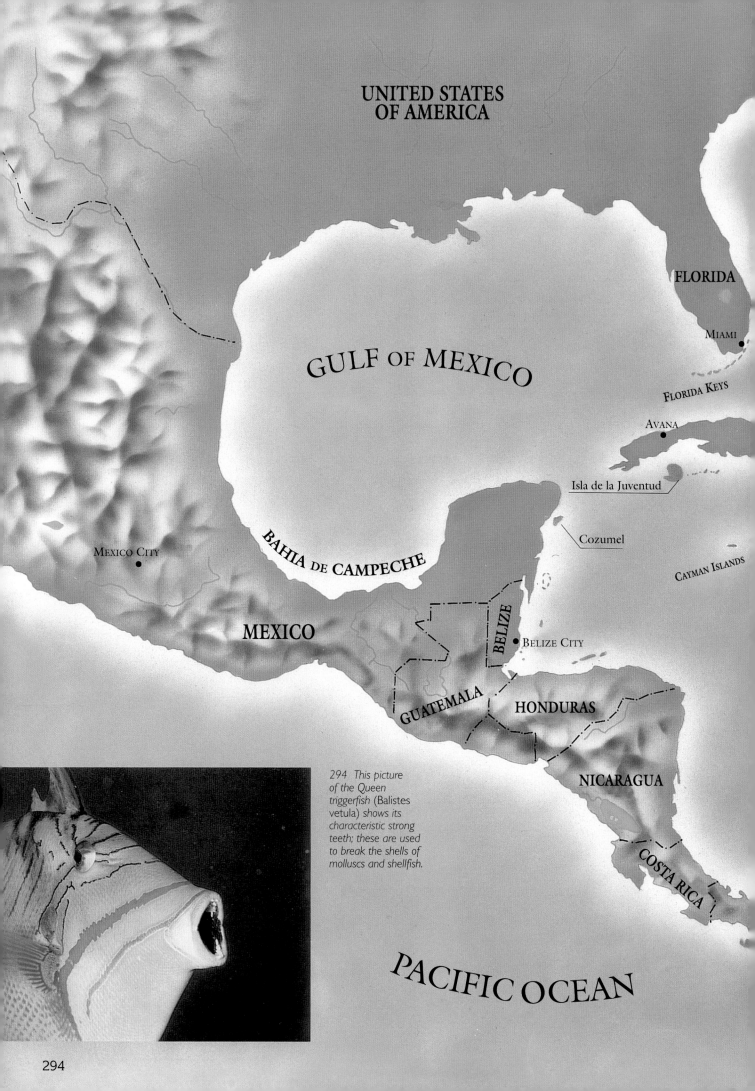

UNITED STATES
OF AMERICA

FLORIDA

MIAMI

GULF OF MEXICO

FLORIDA KEYS

AVANA

Isla de la Juventud

Cozumel

MEXICO CITY

BAHIA DE CAMPECHE

CAYMAN ISLANDS

MEXICO

BELIZE

Belize City

GUATEMALA

HONDURAS

NICARAGUA

294 This picture
of the Queen
triggerfish (Balistes
vetula) shows its
characteristic strong
teeth; these are used
to break the shells of
molluscs and shellfish.

COSTA RICA

PACIFIC OCEAN

# The Caribbean

**Texts**
Angelo Mojetta

**Translations**
Barbara Fisher

## Contents

| | |
|---|---|
| INTRODUCTION | page 296 |
| - THE COASTAL REEFS | page 306 |
| - THE LAGOONS AND THE UNDERWATER PRAIRIES | page 320 |
| - THE PLATFORM REEFS | page 330 |
| - THE EXTERNAL REEFS | page 342 |
| - THE COASTAL WATERS | page 354 |
| - THE GORGONIAN FORESTS | page 360 |
| - THE SPONGES | page 372 |

N

ATLANTIC OCEAN

GRAND BAHAMA

BAHAMAS

NASSAU

Andros

TURKS AND CAICOS

CUBA

INAGUA

DOMINICAN REPUBLIC

SAN JUAN

VIRGIN ISLANDS

Anguilla

HAITI

Saint Cristhopher Nevis

PUERTO RICO

Barbuda

Antigua

Montserrat

Guadeloupe

JAMAICA

SANTO DOMINGO

Dominica

Martinica

Saint Lucia

Saint Vincent

LEEWARD ISLANDS

WINDWARD ISLANDS

Barbados

DUTCH ANTILLES

Grenada

CARIBBEAN SEA

Aruba

Bonaire

Isla de Margarita

Tobago

Curaçao

Trinidad

CARACAS

PANAMA

COLOMBIA

# INTRODUCTION

## THE CARIBBEAN

The area generally known as the Caribbean is like a patchwork of microcosms and seas; these extend along the tropical coastlines of the American continent and then continue in a long curve of islands scattered on the boundary between the Caribbean Sea, internally, and the Atlantic Ocean, externally. This band measures approximately 800 kilometers in width and is nearly 2,000 long, a small surface compared with the expanses of the Pacific Ocean where the Great Barrier Reef of Australia alone occupies only a slightly smaller area. Yet this is one of the richest coral zones, perhaps the only one worthy of this name, in the Atlantic, where more than 600 species of fish and almost a hundred of different hard corals are to be found, most of relatively recent origin (15-10 million years ago) when the lands that make up today's Central America emerged and closed all communication with the Pacific. These and other events that affected the Caribbean region certainly reduced the number of species and genera but at the same time permitted the

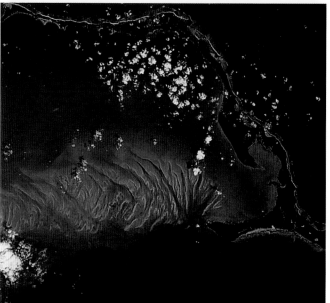

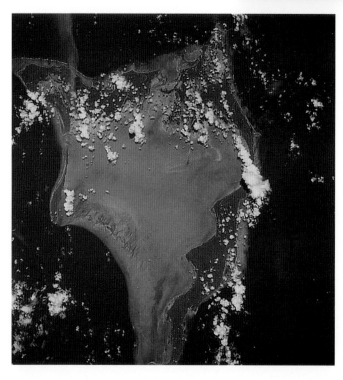

296 top This satellite photograph shows the Florida peninsula partly closing the Caribbean curve to the north. The Bahamas are on the right and lower down is the large island of Cuba which protects much of the Caribbean Sea.

296 center Many of the Caribbean islands - this is Eleuthera Island, Bahamas - have to some degree been shaped by the sea. In certain cases they look as if they are trying to curve backwards and protect large stretches of shallow sea, thus creating the ideal conditions for the settlement and growth of the corals.

296 bottom This photograph taken from the Shuttle shows the Bahamas and Acklins Island in particular. The light patches are shallow waters that may be less than 3 meters deep; the darker areas are abysses plunging to 2,000 meters.

evolution of unusual populations. The Caribbean is dominated by the Caribbean current which flows from east to west and by a coastal counter-current; these influence each other and create numerous internal vortexes. This system of currents is of great importance for the aquatic life as it carries a myriad of larvae, of the most varied organisms from one extreme to the other of the region, contributing to their propagation; this is possible because many of them have a long larval phase. As a result of this movement all the sea beds in the Caribbean present relatively uniform characteristics; from the very first dives they seem dominated in quantity by sponges and gorgonians, taking

very well developed; more than 250 kilometers long, they form what is considered the second largest coral reef in the world after the Great Barrier Reef of Australia. A typical phenomenon of this area are the so-called "blue holes", the remains of underwater caves in which the roof has collapsed leaving intense blue chasms. In contrast the beds of the Lesser Antilles facing the Atlantic are far poorer.
The reefs of Jamaica and the Cayman Islands are limited in size but spectacular, rising at the edge of great depths and often plunging almost vertically for more than 50 meters.

*297 left*
*The positioning of the coral reefs along the islands in the Caribbean Atlantic is influenced by the coastline and exposure to wind and waves. These forces together affect the growth of the corals.*

*297 top right  Only a bird's eye view clearly shows the maze of channels created by nature in the millions of years that have passed since the creation of the Caribbean Sea.*

*297 center right The Caribbean coral reefs are characteristic fringing reefs where corals grow almost parallel to the coast, forming internal lagoons connected to the sea by parallel channels.*

*297 bottom right The elongated development of the reef is favoured by the fact that it spreads over long stretches of shallow rocky areas; here the corals find the ideal conditions to settle and grow.*

the place of the alcyonarians of the Red Sea or the Indo-Pacific. The horizontal expansion of the Caribbean coral formations towards the open sea is limited by the fact that, in just a few kilometers, the sea bed plunges to a depth of 2,000 meters; for this reason they belong mainly to the so-called fringing barriers and mark the coasts from Florida to the Bahamas and as far as Venezuela. A vast platform surrounded by deep sea stretches east of Florida and extends to the Bahamas; it is made up of more than 3,000 islands, cays and rocky banks that create a long barrier, one of the largest in the Atlantic. Andros is of great interest for its unusual geology which has allowed the both sea and fresh waters to open countless passages in the rocks. These caves and blue holes are visited every year by thousands of divers drawn also by the existence of a 228 kilometers barrier, the third longest in the world, which rises less than a mile from the coast. The coral reefs continue with the long curve of islands that stretch from Cuba to Aruba with well-developed fringing reefs thanks to particularly sheltered coasts. Curiously, north of the eastern tip of Cuba there is a small atoll, Reef Hogsty, a totally anomalous coral formation for the Atlantic. The hard coral formations fronting Belize are

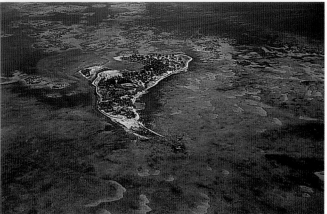

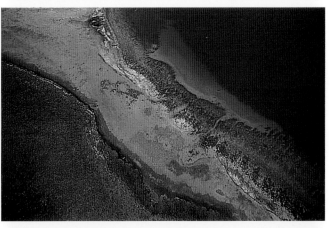

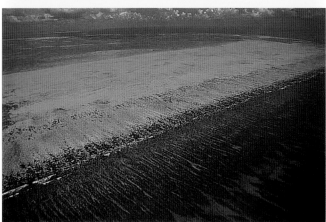

## NATURE ON THE CARIBBEAN COASTLINES

If you explore the coasts of the Caribbean region or even one of the many islands that dot the sea, you will find that the underwater life here is the fruit of the complex interaction of three different ecosystems: the mangroves, the underwater prairies and the coral reefs. Although the first two ecosystems are little visited by divers, experts are well aware of the importance they hold for the coral world, to which they are bound by an extraordinary productivity. The lush mangrove forests, extraordinary plants that manage to colonize the edges of

the sea, take root close to the estuaries and lagoons that form at the mouths of rivers. Here they create a dense barrier that blocks the sediment carried by water, preventing it from ending up in the sea where it could suffocate the corals. Once past the protective mangrove belt, the sea depths are marked by the presence of large prairies capable of exploiting the nutrients that emerge from the mangrove forests with the waters of the swollen rivers and the most abundant tides. Although both these ecosystems are closely bound to the land, the corals, the third vital belt of the Caribbean coasts, are practically self-sufficient, despite the relative poverty of the clear waters where

*298 top left*
*Many points of the emerging Caribbean coast are edged with thick mangrove forests. Extraordinary environments full of life and surprises for those wanting to explore them form on the boundary between land and sea.*

*298 left center top*
*The shallow waters and the channels that mark the entrance to internal lagoons are usually areas of passage for the dense shoals of fish which exploit the tidal current for their movements.*

*298 left center bottom*
*The Atlantic spotted dolphins (Stenella plagiodon) are no rare encounter when sailing from one island to another. Often they approach divers of their own accord and remain with them for some time.*

*298 bottom left*
*A large octopus camouflages itself on the sea bed in a typically defensive position with its tentacles contracted.*

*298 right  The large colonies of elkhorn corals (Acropora palmata) become an attraction for mixed shoals of snappers and grunts. These species love to spend the day in the shade of these hard corals.*

*299 top left*
*The purple outlines of the young Creole wrasse (Clepticus parrae) are frequently seen along the external edges of the reef, where they station in open waters to feed on plankton.*

*299 center left*
*The formations of candelabra sponges rise at intervals on the flattest sea beds, often close to the points where it plunges down into the blue.*

they grow. However different the three ecosystems may be they are in many cases visited and exploited by the same species. The shoals of young grunts penetrate the prairies by night to feed, returning to the corals by day. But this is not all. The French grunts *(Haemulon flavolineatum)* prefer to reproduce in the prairies whereas the barracuda choose the mangroves where the young of many other species, from the butterflyfish to the grouper and parrotfish, also take refuge.

*299 bottom left*
*Filefish (Cantherines sp.) characterized by a small first dorsal fin that has been reduced to a sturdy, spinous ray, are quite curious and will allow a diver to come close if he swims slowly.*

*299 top right*
*This remarkable row pore rope sponge (Aplysina cauliformis) is most common along the deep slopes of the reef.*

*299 bottom right*
*The coney (Epinephelus fulvus) is common in the Caribbean area but it is still not easy to identify; this grouper comes with various markings, including the two-color one seen here.*

# THE CORALS
## OF THE CARIBBEAN

*300 top left  A large colony of giant brain corals* (Colpophyllia natans). *By night the reliefs created by the coral walls are partly hidden by the tentacles of the polyps, expanded in search of food.*

*300 bottom left Intricately branched colonies of lace corals* (Stylaster sp.) *are one of the most striking forms of life in the Caribbean sea. Although similar to fire corals, these do not sting.*

*300 top right A very curious gorgonian species found in the Caribbean is the so-called sea rod* (Plexaura sp.). *The colonies consist in long massive branches attached to a single base.*

*300 bottom right Gorgonian sea fans* (Gorgonia sp.) *abound on the Caribbean reefs. Although most common closer to the surface, where they exploit the currents created by the waves, they are also found at great depths.*

Swimming along a reef in the Caribbean you will encounter a number of areas that differ greatly but which recur fairly regularly. In the waters closest to the surface where the corals are battered by the waves and affected by the flows of the tides there is an abundance of brain corals; these irregular rounded colonies, the surface crossed by twisting growths, cover vast areas of the sea bed. Proceeding towards the open sea there is an increasing profusion of the characteristic colonies of elkhorn acropores *(Acropora palmata),* thus named because their branches are characteristically flattened, like the horns of the large American mammals.
This area presents considerable turbulence as is shown by the many fragments of hard corals that carpet the sea bed. In many cases these manage to reattach themselves and continue to grow. Otherwise they become the substrata on which coral algae settle, serving as a living concrete stabilizing the sea bed and permitting the settlement of smaller corals.
As the waters become deeper (4-10 meters) other bush corals appear but with thin, pointed branches like the horns of a stag *(Acropora cervicornis).*

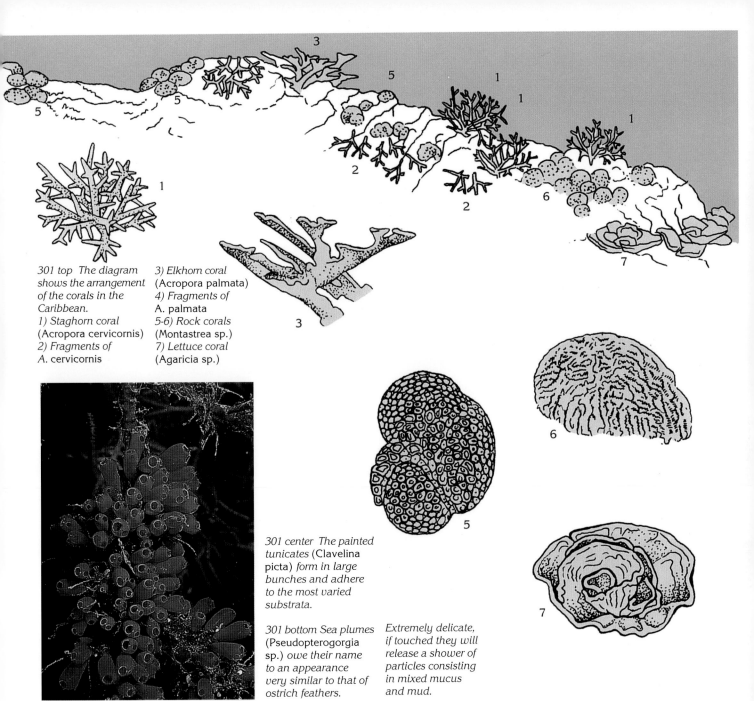

*301 top The diagram shows the arrangement of the corals in the Caribbean.*
*1) Staghorn coral (Acropora cervicornis)*
*2) Fragments of A. cervicornis*
*3) Elkhorn coral (Acropora palmata)*
*4) Fragments of A. palmata*
*5-6) Rock corals (Montastrea sp.)*
*7) Lettuce coral (Agaricia sp.)*

*301 center The painted tunicates (Clavelina picta) form in large bunches and adhere to the most varied substrata.*

*301 bottom Sea plumes (Pseudopterogorgia sp.) owe their name to an appearance very similar to that of ostrich feathers.*

*Extremely delicate, if touched they will release a shower of particles consisting in mixed mucus and mud.*

These corals are more delicate than others and do not become dominant like the former ones. Here you will find gorgonians, fire corals and massive or encrusting hard corals, a prelude to the so-called star corals *(Montastrea sp.)* which make up a large part of the deep reefs in the Caribbean.

# REEF LIFE

Diving on a Caribbean reef will bring you face to face with a strange underwater world: the corals, not so different in shape from those of the Indo-Pacific, are, as already mentioned, flanked by huge sponges of the most incredible shapes and colors. Some resembling huge elephant ears, others barrels, stove pipes or elegant candelabra, pink, red or yellow, they grow so large that a diver can be concealed from the sight of his companion. Given their size it is no surprise that they

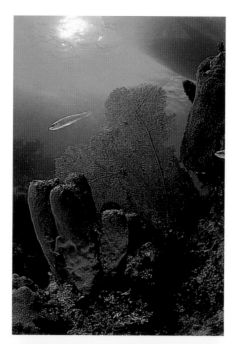

*302 left  The azure vase sponge (Callyspongia plicifera) is one of the most characteristic species in the Caribbean. Despite its common name, the most frequent coloring is pink with fluorescent azure shading at the tip of the opening.*

*302 top right A group of large tube sponges (Agelas sp.) rises from the sea bottom. These favour protected areas and reefs such as those inside canyons or close to caves.*

*302 center right The large eye toadfish (Batrachoides gilberti) is recognized by its ragged barbels and the appendixes around its snout. Typical of sandy, shady bottoms, it tends to remain immobile, partially burying itself in sediments.*

*302 bottom right The delicate Christmas tree worms (Spirobranchus giganteus) are usually seen peeking out from hard coral colonies.*

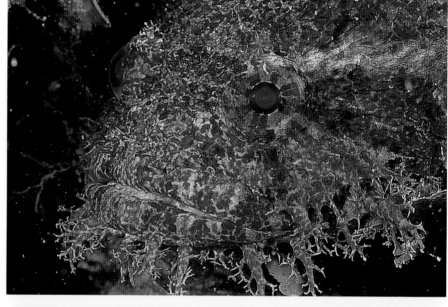

become a sort of microcosm in themselves, surrounded by swarms of wrasse, and damselfish; inside you will often find brittle stars, anemones, worms, shellfish and fish (gobies and blennies) all using the large oscula as dens.
In the shadows of the sponges, lying in wait, are trumpetfish, scorpionfish or the brightly colored seabass of the *Gramma loreto* species, easily recognized by their fluorescent yellow and fuchsia markings.
As well as the sponges, rising from below the walls of the Caribbean reefs are copious forests of gorgonians, from those with regularly woven branches

*303 top left*
*A close up of this roughback batfish (Ogcocephalus parvus) highlights the bizarre form of this creature. These fish usually remain immobile on the sea bed; they rely on their camouflage coloring for safety and will not move even if approached persistently.*

(the Venus sea fans or *Gorgonia flabellum*) to the much larger and irregular *Gorgonia ventalina* which can grow to two meters in height and are mixed with the fragile, plumed, black coral structures, increasingly predominant as the depth increases. Along the walls that depending on the position of the reef, open either towards deep channels or the open sea, the geometry of the sea bed appears intricate and full of holes and crevices abounding with spotted, striped or green moray. The latter, particularly numerous in the Caribbean, have the little appealing but totally undeserved name of *Gymnothorax funebris*.

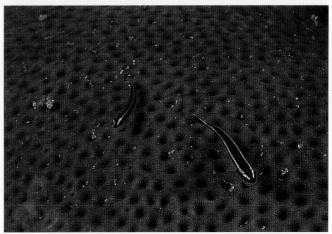

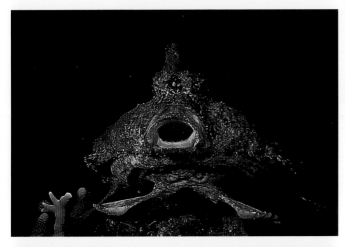

There is no shortage of groupers in the dens, from red *(Cephalopholis fulva)* to striped ones (though this depends on mood as they can change color quickly) of the *Epinephelus striatus* species, better known as the Nassau grouper, and the gigantic *Epinephelus itajara*, the undisputed ladies of the reef and submerged wrecks, visible from afar thanks to the transparent waters and, above all, their size which can considerably exceed two meters. Also lovers of the shadows but less fearsome are the squirrelfish and soldierfish backed by dense shoals of glass fish that reflect flashlight inside the caves.

The groups of fish in this part of the tropical Atlantic are fairly familiar to divers who experienced other coral environments, but the opportunities to make interesting encounters and comparisons are not lacking. The slowly advancing pairs of French angelfish *(Pomacanthus paru)* or grey angelfish *(Pomacanthus*

*303 bottom left*
*The delicate, sinuous, yellow feathered arms of the golden crinoid (Davidaster rubiginosa) stretch out in the current* to capture as much plankton and particles of detritus as possible to feed on. The small body of the crinoid remains hidden in the substratum.

*303 top right*
*The green moray (Gymnothorax funebris) is the largest species in the Caribbean and can grow to 2.4 meters in length. Despite its size it is placid with divers and will allow them to observe it in its den.*

*303 bottom right*
*Fluorescent markings have quite rightly earned this small fish the name of neon goby (Gobiosoma oceanops). Its presence indicates the existence of a cleaning station. This small fish usually spends its time cleaning other fish of parasites and detritus.*

304 top left
A shoal of grey
snappers (Lutjanus
griseus) *circles
slowly above the
flat reef following
the current.
This species often
approaches the
coastline and piers.*

*arcuatus),* up to 50 centimeters
in length, cannot but fascinate,
all the more so because the
latter species seems not to fear
divers and approaches them
of its own accord.
The relaxed movements of these
fish contrasts with the rapid
darting of butterflyfish such
as the *Chaetodon capistratus,*
perhaps the most common
in this area, recognized by
the ocellar spot close to the tail;
young specimens have two,
earning them the name of four-

eyed butterflyfish. Being a lover
of deep reefs, the longsnout
butterflyfish *(Chaetodon
aculeatus)* is very good at
swimming in the thinnest cracks
in the corals in search of
invertebrates.
Constantly in search of food,
brightly colored, curious and
marked by a characteristic jerky
movement are the *Labridae,*
seen in large numbers on any
reef chosen for dive.
For its unusual shape and
unkempt appearance - produced

304 center left
*Keeping an eye
on the diver, this
black grouper*
(Mycteroperca
bonaci) *slowly
swims away to
maintain a safe
distance.*

304 bottom left
*This turtle* (Caretta
caretta) *is often
seen resting in
some sheltered part
of the reef on the
sea bottom.*

by the long rays of the dorsal
fin - it is not difficult to
distinguish the members of
the *Lachnolaimus maximus*
species (tufted hogfish) or the
*Thalassoma bifasciatum* which,
in the early hours of the
afternoon, court each other
close to the surface.
However, not all the fish in
the Caribbean are seen alone
or in pairs. Shoal fish such as
the many *Lutjanidae* or the ever-
present French grunts gathering

304 right
*The silhouette of
the grey angelfish*
(Pomacanthus
arcuatus) *soon
becomes a normal
sight for those
diving on the
northernmost
Caribbean reefs.*

because many species gather at the roots of the trees to reproduce.

You may, amidst the multicolored sponges, oysters, crabs and jellyfish swimming upside down, see small barracuda training for a future in far more dangerous waters, even for them, by following the young of numerous other species from angelfish to parrotfish, or butterflyfish and damselfish.

*305 center right*
*Large silvery tarpons (Megalops atlanitcus) are ideal subjects for photographs thanks to their size and the fact that they live in shoals.*

*305 bottom right*
*The bluestriped lizardfish (Synodus saurus) is known for its skilful camouflage and sedentary habits.*

*305 top left  A small burrfish (Diodon holacanthus) withdraws into the den chosen as a refuge. Very fearful, this fish always moves close to the sea bed trying to camouflage itself.*

*305 bottom left
The longspine squirrelfish (Holocentrus rufus) is not fond of the light and moves slowly by day, trying to keep to poorly illuminated areas.*

*305 top right
Nassau groupers (Epinephelus striatus) are fairly common in the Caribbean where they are much fished. During the mating season they migrate long distances to meet in hundreds and reproduce in certain coastal areas.*

in hundreds to form bright yellow and blue processions, or the large tarpons, more than two meters long will alone make even the exploration of an old landing stage unforgettable or become a landmark for a dive thanks to their stable habits. Ever present are walls of silver jacks and the streaking schools of barracuda that may suddenly break open to leave a passage for sharks, hammer sharks included, rays, eagle rays or large stingrays that animate the sandy beds, more populated than would be imagined in these seas where the alternation of corals, rocks, sand and prairies is often seen.

Lastly, a little advice for those who love to explore.

Between one dive and another, if you are in the vicinity, do not fail to visit a lagoon surrounded by mangroves.

This particular environment is halfway between dry land and sea and is of considerable importance for the Caribbean

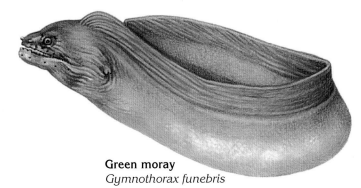

**Green moray**
*Gymnothorax funebris*

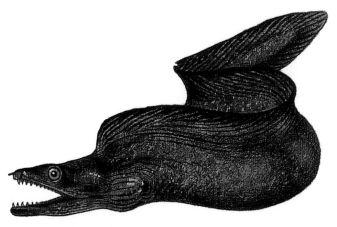

**Purplemouth moray**
*Gymnothorax vicinus*

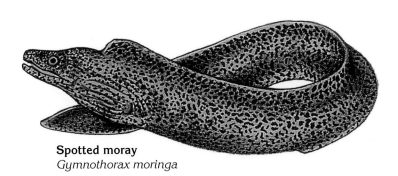

**Spotted moray**
*Gymnothorax moringa*

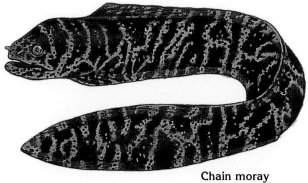

**Chain moray**
*Echidna catenata*

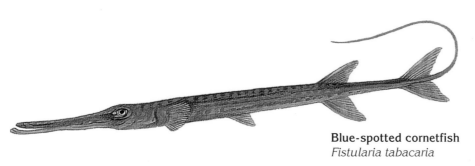

**Blue-spotted cornetfish**
*Fistularia tabacaria*

*306-307 This large moray has been caught in the open water beside a grouper - a rather unusual meeting, especially by day.*

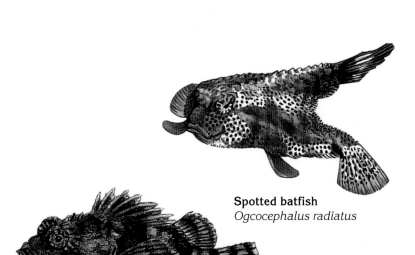

**Spotted batfish**
*Ogcocephalus radiatus*

**Yellowtail damselfish**
*Microspathodon chrysurus*

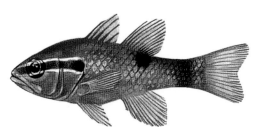

**Coral scorpionfish**
*Scorpaena plumieri*

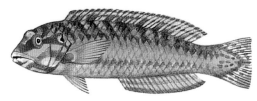

**Slippery dick**
*Halichoeres bivitattus*

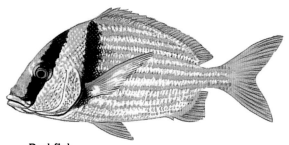

**Spotted cardinalfish**
*Apogon maculatus*

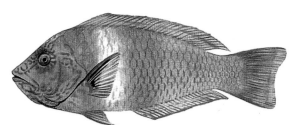

**Puddingwife**
*Halichoeres radiatus*

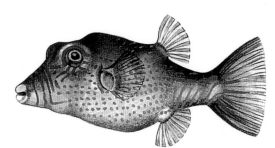

**Porkfish**
*Anisotremus virginicus*

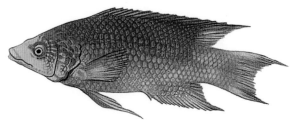

**Spanish hogfish**
*Bodianus rufus*

**Longnose pufferfish**
*Canthigaster rostrata*

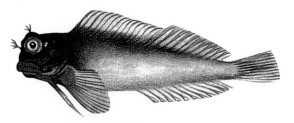

**Red-mouthed blenny**
*Ophioblennius atlanticus*

**Sergeant major**
*Abudefduf sexatilis*

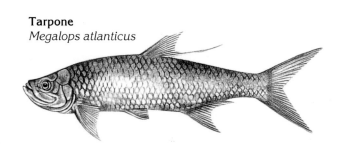

Tarpone
*Megalops atlanticus*

*308-309 Sergeant majors (Abudefduf saxatilis) are quite common along the reefs where they tend to gather in small shoals.*

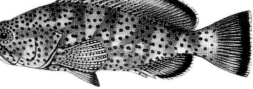

**Red hind**
*Epinephelus guttatus*

**Yellowhead jawfish**
*Opistognathus aurifrons*

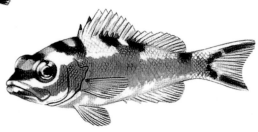

**Tobaccofish**
*Serranus tabacarius*

309

The rocky coasts of the tropical band of the American continent and the Caribbean islands are an ideal habitat for the growth of numerous species of hard corals and other encrusting and building organisms. The constant development of these creatures leads to the formation of characteristic fringing barriers that come between the coast and the open sea, towards which they continue to grow until depth or the lack of a rocky substratum make it impossible for them to live. The short distance that separates them from the coast makes this type of sea bed easy to explore even for those who snorkel; you can pass rapidly from a warm sunny beach to the transparent waters of a reef reached along one of those tongues of sand that often cross them.

Swimming along one of these paths, accompanied by an odd wrasse or bass, you will realize that you are approaching a reef thanks to the gradual appearance of scattered blocks of coral which, established on some submerged rocky spur, never emerge at low tide.

The influence of the sea on this type of reef is very clear. Its upward growth is limited by the tides and its outline and the arrangement of the corals are affected by the waves which often break with considerable force, especially during the frequent hurricanes.

*310-311 Elkhorn corals (Acropora palmata) are among the principal reef builders in the Caribbean. They are more numerous in fringing reef areas where the hydrodynamism is more intense.*

*311 top Large blocks of hard corals quickly become small microcosms full of life on which sponges, gorgonians, corals and truncates settle.*

*311 bottom Pillar corals (Dendrogyra cylindrus) are among the few corals whose polyps are visible during the day. The colonies can rise to three meters in height.*

*312 top*
*This nudibranch has been photographed while feeding on a large sponge.*

*312 bottom*
*The crevices and caves at the base of coral reefs are the ideal habitat for Caribbean spiny lobsters (Panulirus argus). They leave their dens during the night to search for food.*

*312-313*
*Some cleaner shrimps such as this* Periclimenes *live in association with large sea anemones. Hidden amidst the tentacles only their long antennae emerge to attract fish in need of cleaning.*

A fringing reef is characterized by a wide variety of coral species, although often these, in the end, take on massive, irregular forms.

Pinnacles up to two meters high follow hills of layers of overlapping corals, almost as if suddenly transformed from liquid to solids. Here the action of the waves excavates large craters in the rocks and the bottoms of these fill with sand and detritus. This hides small shellfish and molluscs and provides personal territory for the damselfish, rich in algae and easily defended against all the other herbivorous fish. As well as these corals, and again swimming in just a few meters of water, the diver will notice the unmistakable contours of the brain corals, covered with the twisting growths that protect the large tentacles of their greenish polyps. Where the waves are less intense star corals grow, thus called for their star-shaped cups which stand out like dark spots on the light-colored surface of the colonies. The fire corals are almost golden with white edging; as in the Indo-Pacific these are present in the upper layers of the reef where they form an unwelcome community, their polyps giving a considerable sting.

*313 bottom left*
*The shellfish family the so-called yellowline arrow crabs belong to are marked by long, slender legs that seem to hold the body suspended in the water.*

*313 bottom right*
*Many forms of life on the sea bed entrust survival to their small dimensions and, above all, camouflage coloring that protects them from attacks by predators.*

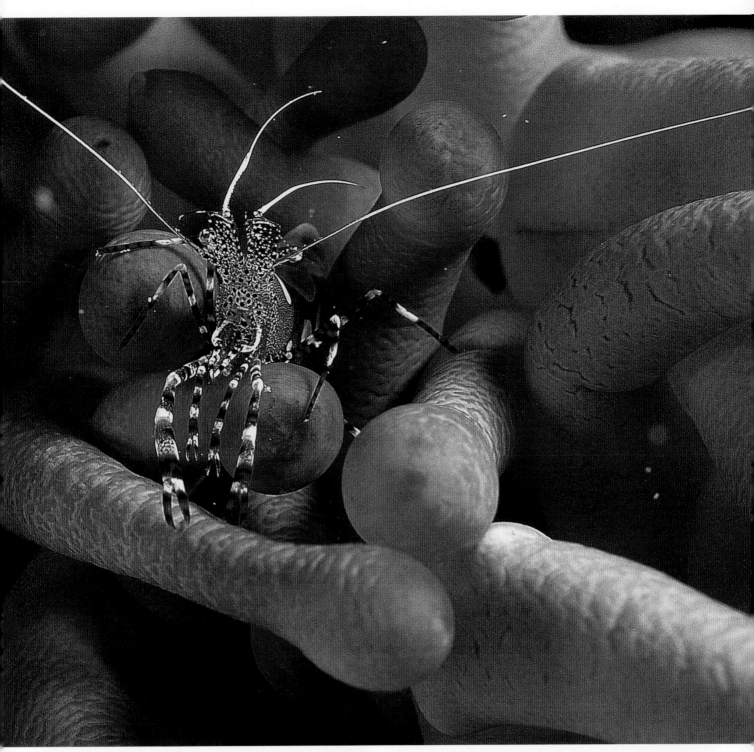

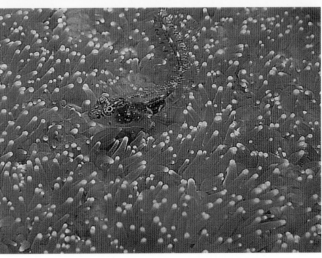

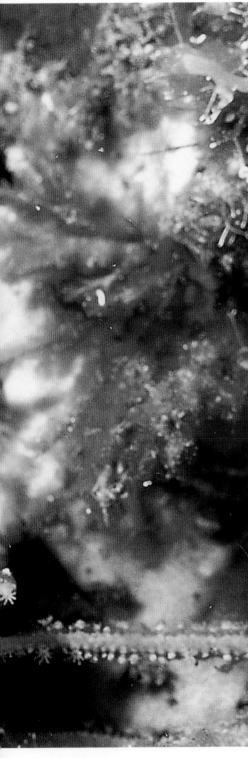

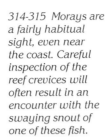

*314-315 Morays are a fairly habitual sight, even near the coast. Careful inspection of the reef crevices will often result in an encounter with the swaying snout of one of these fish.*

*315 The spotted moray (Gymnothorax moringa) lives along the shallower reefs and even in the eel-grass prairies.*

*314 bottom left The gaping mouth of the moray is not a manifestation of aggressivity but a need bound to the way this fish breathes.*

*314 bottom right Small yellow spots on the snout are the distinctive feature of this unusual sharptail eel (Myrichthys breviceps).*

Perhaps the most fascinating part of this coastal reef is in the depths.
Here live the elkhorn hard coral colonies which you will be tempted to enter at the risk of finding yourself enclosed in countless long branches and from which it is hard to emerge without damaging the corals.
It is best to avoid such temptations especially because, even at a safe distance from the bottom, the sight is fantastic. Where the caves and crevices open morays peek out, including giant green ones with a scientific name - *Gymnothorax funebris* - that seems chosen purposely to perpetrate the unjustified fear that surrounds these fish.

316-317 Shoals
of large tarpons
(Megalops
atlanticus) *move*
*mostly by night*
*and tend to station*
*in the shelter*
*of canyons and*
*sheltered parts*
*of the reef by day.*

*316 top  It is hard*
*to tell if the*
*photographer*
*took a picture*
*of a sponge or of a*
*scorpionfish, well*
*camouflated and*
*waiting for a prey.*

*316 center*
*The sandy beds that*
*form at the base of*
*the reef caves provide*
*shelter for large*
*spotted scorpionfish*
*(Scorpaena*
*plumieri).*

*316 bottom*
*A close-up of a*
*spotted scorpionfish*
*shows its*
*characteristic, wide*
*mouth surrounded*
*by ragged, fleshy*
*growths.*

Groupers are not lacking and the same can be said of cardinal fish, not lovers of light, but by no means invisible to those who peer into a crack in the reef. In the quieter areas, perhaps close to an old landing stage or a large break in the reef, you will catch sight of large silver silhouettes; on first sight, these could be mistaken for barracuda but in actual fact are tarpons. These fish love to gather by day and scatter at night to hunt.

318-319 *The young of the stoplight parrotfish (*Sparisoma viride*) do not pass unnoticed, thanks to their bright red coloring.*

318 bottom *A red hind (*Epinephelus guttatus*) swims close to the reef. Despite its apparently bright coloring, this fish camouflages very well underwater.*

319 top *Cardinalfish are commonly found in poorly lighted areas. Their large eyes are an adaptation to the reduced light.*

319 bottom *A yellowhead jawfish (*Opistognathus aurifrons*) swims above a den dug in the sea bed. At the slightest sign of danger it will dive into it, disappearing immediately from sight.*

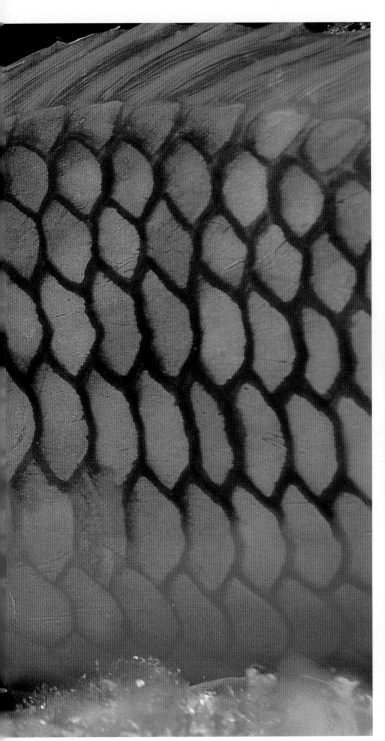

hesitation approach what, despite its unusual shape, should be considered a fearful predator. Well-defended against these attacks is the small *Canthigaster*, a pufferfish that combines its ability to swell with flesh that is poisonous for all predators.

The fringing reef often ends in the sand, rarely devoid of life as is shown by the dens of many jawfishes; these appear to be holes of no interest but, with just a little patience, you will see elongated fish with a yellow head and blue body, peek out seemingly from nowhere; once reassured they decide to emerge but only partially, in search of plankton.

The forests of elkhorn corals are interwoven, often in large numbers, with fans of the gorgonians, destined to become a recurrent sight on all dives in the Caribbean.

Where all these organisms mix the underwater life becomes incredibly lush.

Groups of striped sergeant-majors hover in the water, mingling with the larger grunts and *Labridae* or small yellowtail damselfish commonly found at the young stage among fire corals.

The gorgonians offer an ideal camouflage for those waiting to spot trumpetfish *(Aulostomus maculatus)* which, in a vertical position, allow themselves to drift with the currents, deceiving the other fish who without

# THE LAGOONS AND THE UNDERWATER PRAIRIES

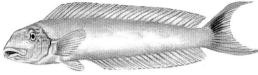

**Bonefish**
*Albula vulpes*

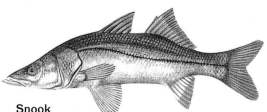

**Snook**
*Centropomus undecimalis*

**Sand tilefish**
*Malacanthus plumieri*

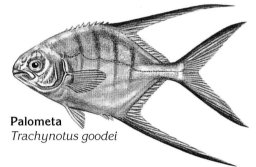

**Palometa**
*Trachynotus goodei*

**Sea-horse**
*Hippocampus erectus*

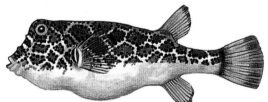

**Chequered pufferfish**
*Sphoeroides testudineus*

**Bandtail pufferfish**
*Sphoeroides spengleri*

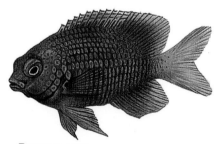

**Beaugregory**
*Stegastes leucostictos*

**Ocellated turbot**
*Bothus lunatus*

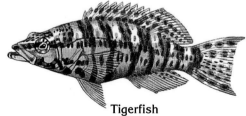

**Tigerfish**
*Serranus tigrinus*

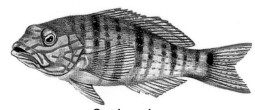

**Sand perch**
*Diplectrum formosum*

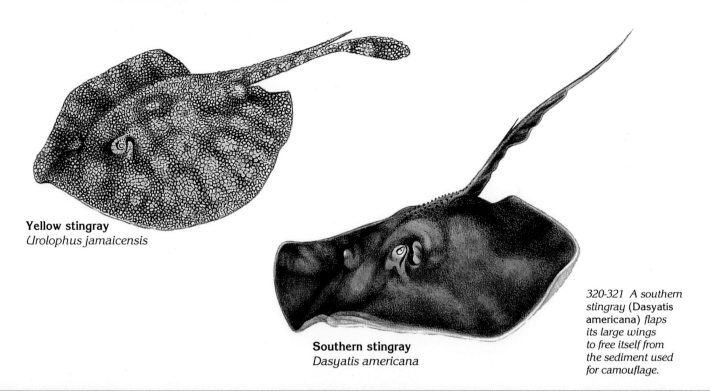

**Yellow stingray**
*Urolophus jamaicensis*

**Southern stingray**
*Dasyatis americana*

*320-321 A southern stingray (Dasyatis americana) flaps its large wings to free itself from the sediment used for camouflage.*

322-323 Abandoned shells soon become prey for hermit crabs which use them as shelter, their bodies being only partially protected by a shell.

322 bottom
The prairies of marine plants and sandy beds are the favourite habitat of the Florida fighting conch (Strombus alatus). As this gasteropod moves on the sea bed it explores the sediments with is long proboscis and tentacles in search of prey.

A wide expanse of clear waters, broken by darker areas seemingly swaying to and fro beneath the waves that trace a white edging farther out. This is the panorama that may appear from the top of a dune or even from the beach and reveals the presence of a lagoon.
Far from being the mere continuation of a stretch of sandy coastline, the lagoon is the ideal place for enjoyable explorations. If the sea bed seems uniform in composition, the forms of life will be greatly varied. Small mounds conceal the flat urchins known as "sand dollars" and other reliefs

*323 These large crabs usually abandon the reef to inspect the expanses of sand in search of molluscs or dead fish to feed on.*

will hide bivalve molluscs and gasteropods. Much more active, but equally hidden or ready to flee if threatened, are the numerous crustaceans, including the large box crabs, very similar to the Mediterranean shame-faced crabs, or the red hermit crabs that live in the empty shells of the queen conch, more similar to armour plating than mollusc niches.
The layers of water above may also hold unexpected surprises sometimes represented by a swimming sea hare or swarms of harmless, pink moon jelly, with their characteristic clover shape.

But the lagoons are not populated by invertebrates alone. There is no shortage of fish and they have different ways of appearing or disappearing.
A vortex of mud in the shallow water from which forked tails stick out signals the presence of white bonefish *(Albula vulpes)* rooting away busily in the sea bed in search of shellfish and molluscs.
A slight distance away, as if created by the foam of the fringing reefs they love to station in, appears a group of silver Palometa *(Trachinotus goodei)*; dark-striped, their fins edged with black, these blend with the sparkling glimmer of the shallow

*324 It is not unusual to see a large grouper immobile on a stretch of sand totally devoid of refuge. But as soon as you get too close it will flee greatly agitating its tail.*

*324-325 Atlantic spadefish* (Chaetodipterus striatus) *are easily recognized by their silvery color and black vertical stripes. They live in small shoals in the open water close to the reef.*

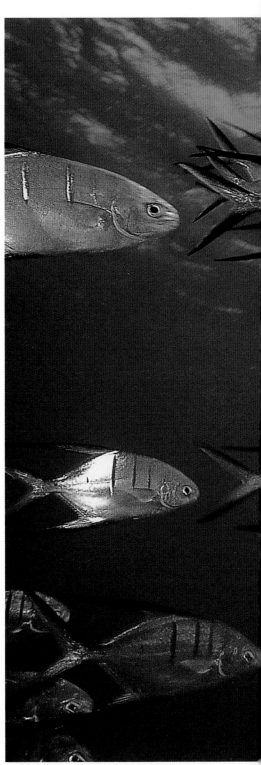

waters; also streaking past here are small shoals of tapered snook *(Centropomus undecimalis)*, with a characteristically elongated snout and concave upper profile. If you follow them they will often lead you to the edge of the lagoon where this borders with the outposts of the reef or the prairies of sea phanerogams; it may be interesting to penetrate these at least for a brief exploration, trying to avoid touching the bottom so as not to sink partially into the fine sediment.

*325 bottom left
At dusk groupers, larger ones in particular, leave their dens to hunt.*

*325 bottom right
This close up of a needle-fish shows the long mouth and numerous teeth that permit it to seize the small fish it feeds on close to the water's surface.*

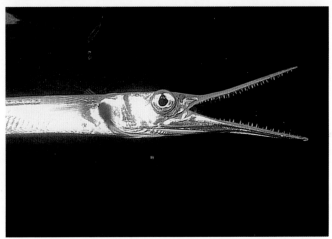

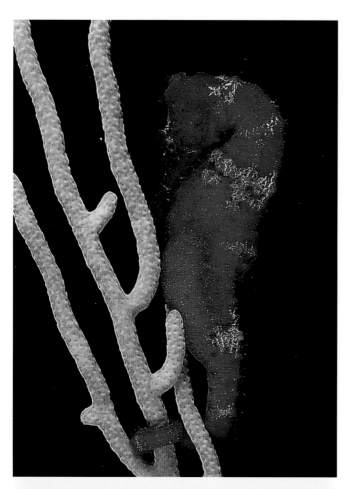

Damselfish, wrasse, small groupers all swim above the fronds amid which are concealed the sea horses that entrust all defence to their shape and color. After this brief foray into the thick prairie you can turn back towards the lagoon, perhaps remaining on the boundary between the two environments, thus seeing the best of both worlds. Curiosity will soon be satisfied.

A strange vertical fin, seemingly emerging isolated from the sand, may signal the presence of an ocellated turbot, with its characteristic blue spots; or close to large holes in the bottom sand perches and sand tilefish can be seen stationary, but ready to dive into their dens if approached too abruptly. This escape will not however hide them from sight because often these dens are almost vertical and, if you peep into the opening, the fish are visible inside them.

This roaming close to the sea bottom could even turn into an involuntary way to draw the pufferfish, capable of swelling up by swallowing water. Anything can happen underwater and sediment moved by your fins may attract these curious fish, ready to take advantage of the invertebrates momentaneously uncovered by those strange, shining and noisy beings that have invaded the sea.

*326 top left*
*This sea horse is firmly fixed to the tip of a gorgonian fan branch.*

*326 bottom left*
*A peacock turbot (Bothus lunatus) tries to camouflage itself on the sea bed but its unmistakable blue spots remain clearly visible.*

*326 right*
*This male sea horse (Hippocampus erectus) can be easily recognized by the ventral slit of the marsupium. The fish sways attached to a gorgonian, because perhaps it is waiting for a female.*

*327 The lined sea horse usually remains in the gorgonians and corals, using its prehensile tail to cling to the branches. In this photograph it remains immobile allowing itself to be cradled by the currents.*

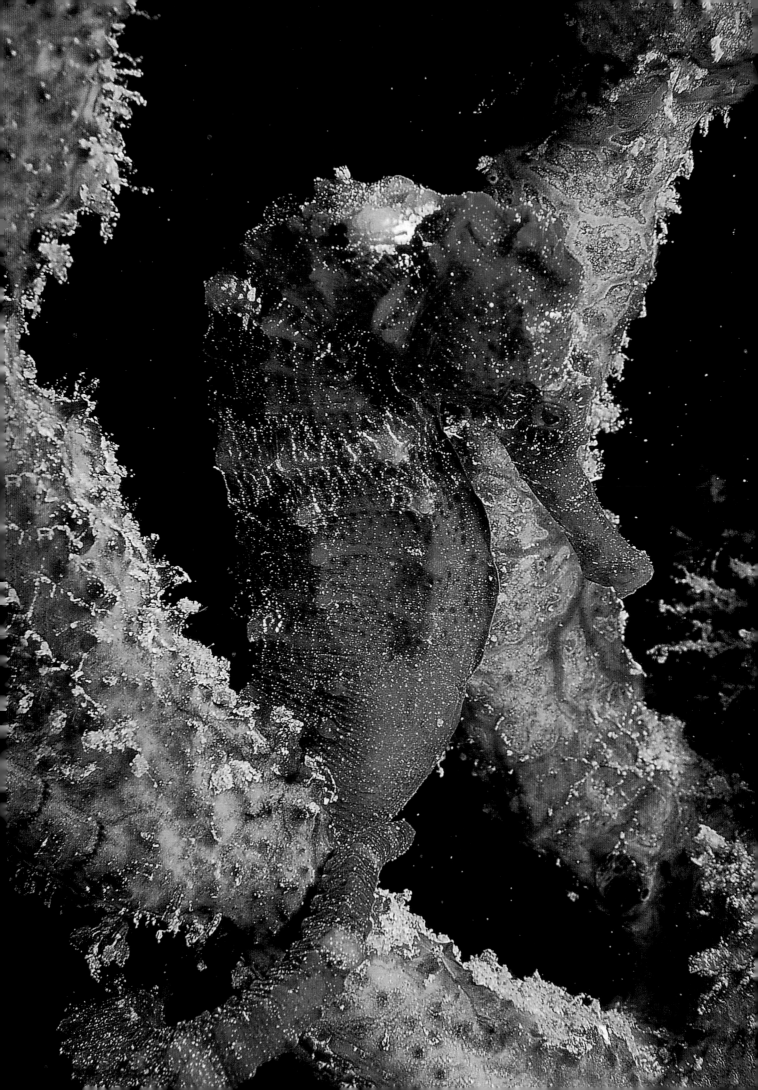

*328 top  A yellow stingray (Urolophus jamaicensis) has stopped on a sandy sea bed and has started to change color for camouflage. Characteristic of the sandy areas around the reefs, it has a stumpy tail with spines.*

*328 center The southern stingrays (Dasyatis americana) have been famous since Stingray City became one of the favourite haunts of divers in the Caribbean. Only here is it possible to swim safely with these creatures.*

*328 bottom The underwater prairies are another habitat popular among stingrays in search of molluscs and shellfish.*

*328-329 This
yellow stingray
(Urolophus
jamaicensis) is
well-camouflaged
on the sea bed.*

The sovereigns of the Caribbean lagoons remain the large stingrays both yellow *(Urolophus jamaicensis)* and southern ones *(Dasyatis americana)*, the most famous since they have had their own city, Stingray City. To visit it you must go to North Sound, a vast lagoon of Grand Cayman. Here hundreds of stingrays swim freely, approaching divers without fear in search of the fish it is now customary to offer. Visits to Stingray City are free but it may be more reassuring to be accompanied by guides for those who harbour some fear of these creatures; their unusual conduct is a unique spectacle that it would be pointless to seek elsewhere at the risk of provoking dangerous reactions in wilder specimens less accustomed to the presence of man.

# THE PLATFORM REEFS

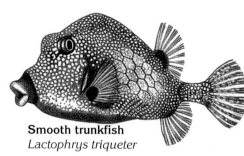

**Smooth trunkfish**
*Lactophrys triqueter*

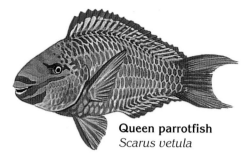

**Queen parrotfish**
*Scarus vetula*

**Red-banded parrotfish**
*Sparisoma aureofrenatum*

**Spotlight parrotfish**
*Sparisoma viride*

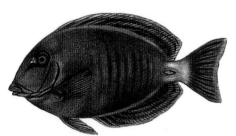

**Surgeonfish**
*Acanthurus chirurgus*

**Bluetang**
*Acanthurus coeruleus*

**Bluehead wrasse**
*Thalassoma bifasciatum*

**Lizardfish**
*Synodus intermedius*

**High hat**
*Pareques acuminatus*

**Nassau grouper**
*Epinephelus striatus*

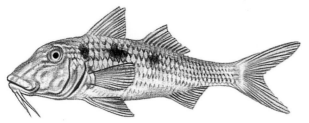

**Triglia maculata**
*Pseudopeneus maculatus*

**Triglia gialla**
*Mulloidichthys martinicus*

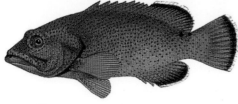

**Coney**
*Epinephelus fulvus*

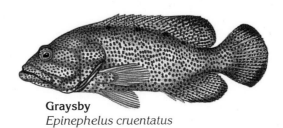

**Graysby**
*Epinephelus cruentatus*

*330-331*
*The honeycomb cowfish (*Lactophrys polygonia) *is easily identified by its reticulate markings and the pointed spine above each eye.*

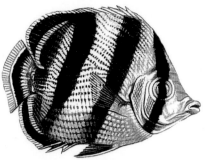

**Chaetodon butterflyfish**
*Chaetodon striatus*

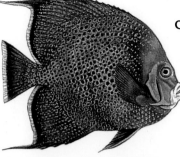

**Grey angelfish**
*Pomacanthus arcuatus*

**Rock beauty**
*Holacanthus tricolor*

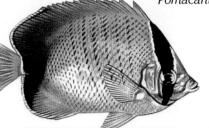

**Reef butterflyfish**
*Chaetodon sedentarius*

**Foureye butterflyfish**
*Chaetodon capistratus*

**Spotfin butterflyfish**
*Chaetodon ocellatus*

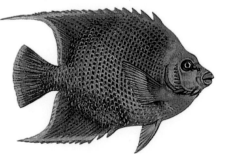

**Blue angelfish**
*Holacanthus bermudensis*

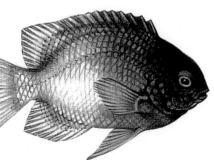

**Bicolor damselfish**
*Stegastes partitus*

**Brown chromis**
*Chromis multilineata*

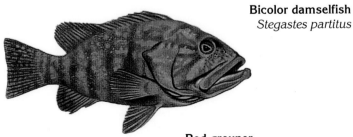

**Red grouper**
*Epinephelus morio*

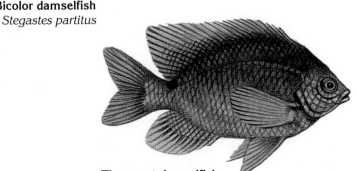

**Threespot damselfish**
*Stegastes planifrons*

The term platform conjures up the image of wide flat surfaces but in the case of the coral reef this word has a different meaning, at least as concerns its extension. A darkening of the waters is a forewarning of the presence of the platform or the crest of the reef, this being a more precise definition as it stresses its meaning of boundary.

Anyone approaching the marine world must learn to recognize the signs it gives and that will constantly prompt different itineraries.

The platform is the upper part of the reef and although its upper limits coincide with those reached by the low tide, its extension depends on how rapidly the rocky bed descends to greater depths. Whatever its size, a platform is recognized by the sharp increase in presence of hard coral species. The corals that form the fringing reef are joined by *Porites*, yellow in color and with a surface riddled with small holes, and other branched corals, ranging from those with short branches to those with long pointed ones such as the staghorns.

*334 top left A star coral monolith rises from the sea bed. These hard corals adopt various forms according to the substratum they settle on.*

*334 bottom left The shape of a sponge depends on the currents, exploited by these creatures for better oxygenation and filtering.*

*334 top right Although the proportions are distorted by the perspective, some huge gorgonian fans can be found in the Caribbean.*

*334 bottom right In the competition between sponges and gorgonians for the substrata it is often the latter who succumb, being literally engulfed in the tissues of the more active sponges.*

*335 Brain corals owe their name to the shape of the polyp calyxes; these unite and intertwine like the convolutions of the brain.*

Of course every species settles in certain points and not others, according to a hierarchy that makes the fire and elkhorn *(Acropora palmata)* corals predominant on the surface; for these the action of the waves is essential to maintain the polyps free from sediments, they themselves being devoid of mechanisms that can eliminate them. On the other hand the action of the waves, especially the more violent ones, can break the fragile branches or even entire colonies creating large areas occupied by coral detritus; this in part crumbles and in part is compacted together by calcareous algae, sponges and other encrusting organisms.

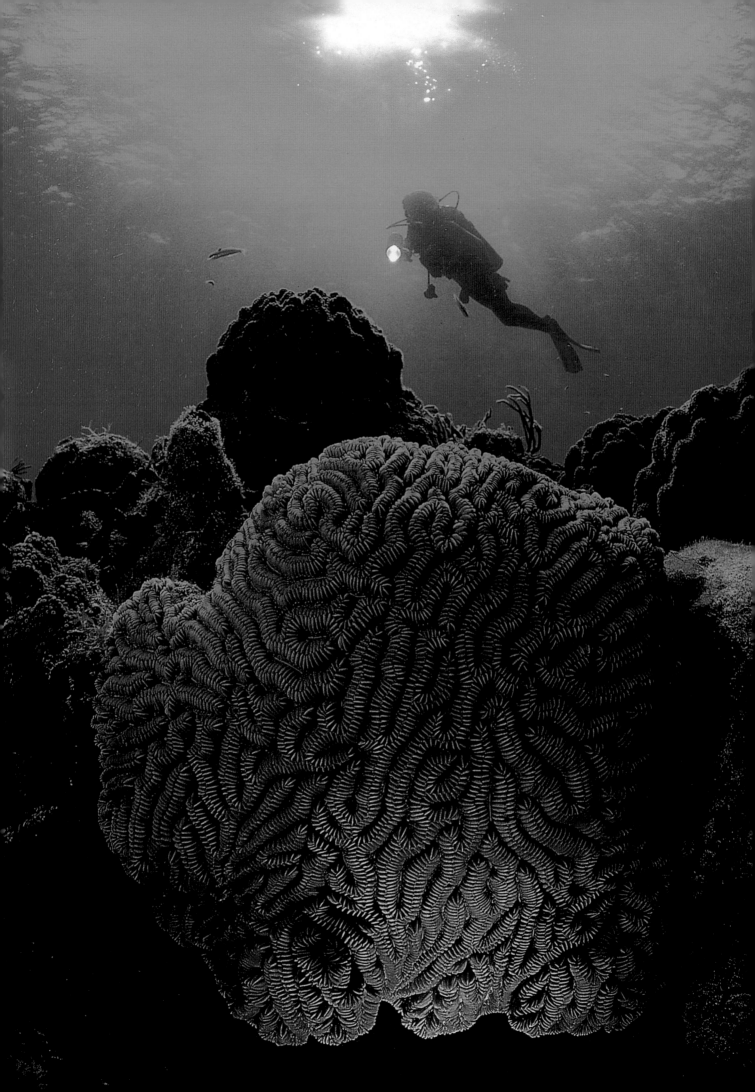

*336 top  A front view of a Nassau grouper shows that it fully deserves the scientific name of* Epinephelus striatus.

*336 bottom A sturdy head and aggressive snout identify groupers as predators without a shadow of doubt. The sharp teeth serve mainly to hold the prey which is then swallowed whole.*

*336-337 This unusual golden coloring is one of the peculiarities in the life of the coney (*Epinephelus fulvus*), quite hard to identify underwater because of the variety of markings that this species can boast.*

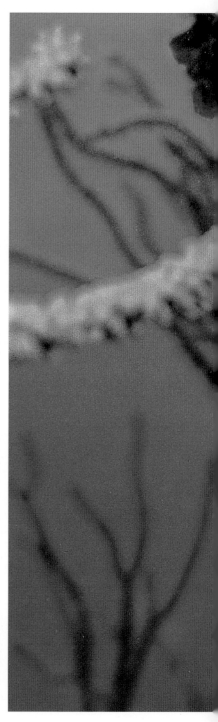

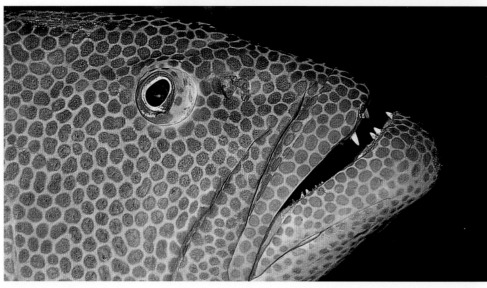

More massive corals appear where the hydrodynamism becomes less accentuated; these form hills or pinnacles, providing the ideal substratum for other hard corals, encrusting and lovers of light which they exploit to the full extending their colonies in a semicircle.

Again, the action of the waves and the currents is obvious. In some points the colonies of hard corals seem to grow as if they had been planted uniformly by man. Elsewhere they develop with a twisted symmetry, farther on canyons and tunnels, carpeted with gorgonian fans and sponges, signal the presence of currents; it is sometimes enjoyable to abandon yourself and effortlessly drift through these labyrinths. Inside, the cracks and caves house striped highhats and various species of grouper, from the red ones to the spotted or Nassau groupers, perhaps the most common in the Caribbean and remarkable not just for the ease with which they change markings but also because, every year, they take part in an impressive mass migration that results in their meeting in thousands along the coasts of Belize to reproduce.

The rays of light that penetrate the canyons sometimes become floodlights and illuminate shoals of blue surgeon fish, *Acanthurus coeruleus* or isolated specimens of their relatives *Acanthurus chirurgus* which station closer to the surface.

This is the favourite haunt of butterflyfish, perhaps less colorful than those in the Indo-Pacific but equally busy swimming through the corals in search of polyps, and angelfish.

The most common of the latter, and not just in this area, is the grey angelfish *(Pomacanthus arcuatus)* which, 50 centimeters in length, dominates all the other members of the family. Smaller but more colorful is the Rock beauty *(Holacanthus tricolor)* with its blue mouth and yellow and black body, usually found alone patrolling an area full of sponges, its favourite food. The young of this species, lemon in color with a black ocellar spot edged with blue on the flanks, immediately draw the attention of divers passing in the upper part of the platform area but it is best to be wary as these little fish like to stay close to

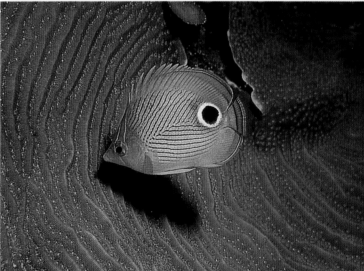

fire corals. Almost as if links in an invisible chain, the shape of the young of the *Holacanthus tricolor* is similar to that of the damselfish of which numerous species exist. Some of these *(Stegastes partitus, Stegastes planifrons)* can prove rather aggressive and demonstrate the control of their territory by facing a diver with their fins spread wide and mouth wide open or attacking directly should there be eggs to defend.

Very different is the attitude of the smooth trunk fish *(Lactophrys triqueter)* which, trusting in the protection offered by their armoured body, allow divers to approach without fear, like the Scaridae or parrotfish that, when busy chewing corals, will let divers watch and photograph them at length before swimming away, more irritated than frightened, with sharp fin strokes. The pieces of coral chewed by the parrotfish rapidly become coral sand which, expelled at regular intervals by the fish, falls on the sea bed to form large sandy areas; the position of these depends on the currents and they become the ideal hunting ground for mullet, both the spotted day ones and the yellow-tailed ones with more nocturnal habits.

*340 top and center*
*The stoplight parrotfish (Sparisoma viride) changes color as it grows. Top a male adult; below the young with its smart red-chequered markings.*

*340 bottom*
*The parrotfish usually feeds on the algae that grow over the corals, breaking them with strong teeth that have developed like the beak of a parrot.*

340-341 The picture shows a spotted goatfish (Pseudopeneus maculatus). *This fish loses it blackish spots on the sides when it remains quietly immobile on the sea bed.*

*341 top The tendentially oval body, dark coloring and falcated tail mean the outline of this surgeonfish is fairly easy to distinguish, even from a distance.*

# THE EXTERNAL REEF

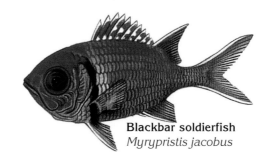

**Blackbar soldierfish**
*Myrypristis jacobus*

**Spotfin hogfish**
*Bodianus pulchellus*

**Glasseye snapper**
*Priacanthus cruentatus*

**Cubera snapper**
*Lutjanus cyanopterus*

**French grunt**
*Haemulon flavolineatus*

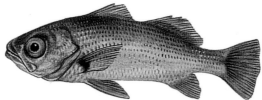

**Reef croaker**
*Odontoscion dentex*

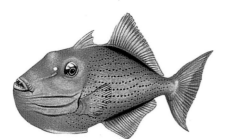

**Sargassum triggerfish**
*Xanthychthys ringens*

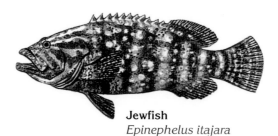

**Jewfish**
*Epinephelus itajara*

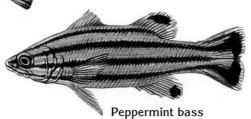

**Tiger grouper**
*Mycteroperca tigris*

**Peppermint bass**
*Liopropoma rubre*

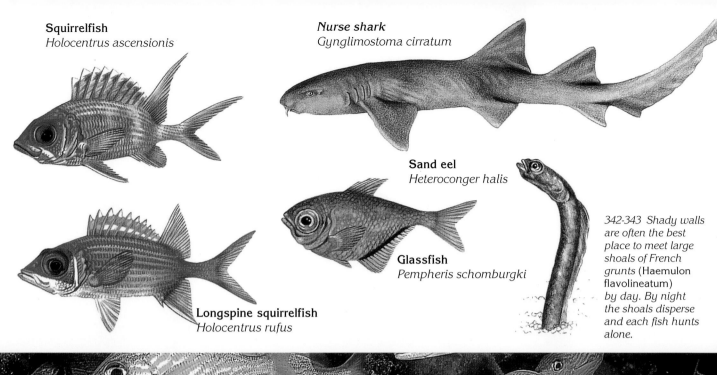

**Squirrelfish**
*Holocentrus ascensionis*

**Nurse shark**
*Gynglimostoma cirratum*

**Sand eel**
*Heteroconger halis*

**Glassfish**
*Pempheris schomburgki*

**Longspine squirrelfish**
*Holocentrus rufus*

342-343 Shady walls are often the best place to meet large shoals of French grunts (Haemulon flavolineatum) by day. By night the shoals disperse and each fish hunts alone.

**Surgeonfish**
*Acanthurus bahianus*

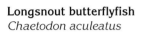

**Longsnout butterflyfish**
*Chaetodon aculeatus*

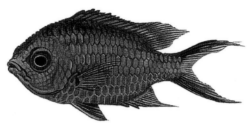

**Blue chromis**
*Chromis cyanea*

**Creole wrasse**
*Clepticus parrae*

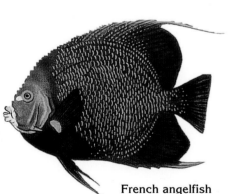

**Cherubfish**
*Centropyge argi*

**French angelfish**
*Pomacanthus paru*

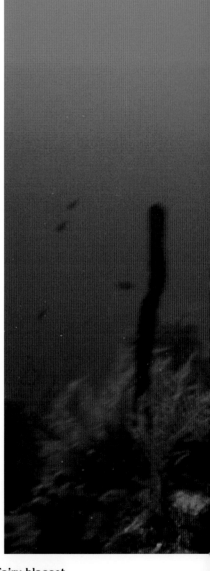

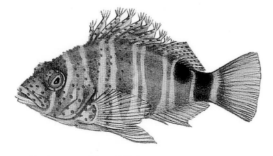

**Redspotted hawkfish**
*Amblycirrhitus pinos*

**Fairy blasset**
*Gramma loreto*

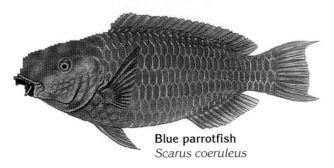

**Blue parrotfish**
*Scarus coeruleus*

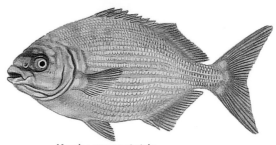

*Kyphosus sectatrix*

**Jack-knife fish**
*Equetus lanceolatus*

344-345
The French angelfish
(Pomacanthus paru)
*is easily recognized
by an eye edged
with yellow and*
backwardly
protracting fins.
*The young have
curious habits,
acting as cleaner
fish.*

345

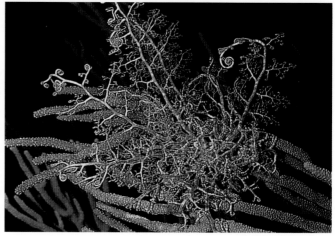

346

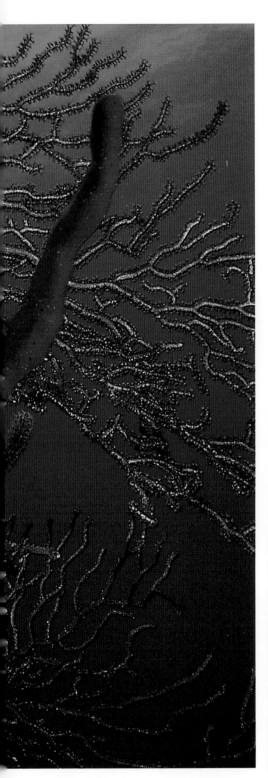

Nearly everywhere the bathymetric ten-meter line marks the boundary that must be crossed to enter the heart of the reef. Tongues of sand broken by rocks and coral formations almost parallel to each other will lead whoever follows them towards the outer edge of the reef, where the slope becomes almost vertical, plunging sometimes towards sea beds hundreds of meters below. The staghorn hard corals form an almost uninterrupted band of colonies, their dimensions becoming ever smaller as the depth increases. This happens also to the other hard corals that mix their column, leaf, rounded mass or pointed pagoda shape colonies; the latter type of formation is quite common in the Caribbean where in some areas it can cover up to half the sea bed. The base and the sides of the corals rapidly become ideal substrata for settlements of gorgonians, sponges, calcareous algae, sea anemones and numerous other invertebrates that add colors and life to these depths.

*347 Deep, vertical walls are the favourite habitat of certain gorgonians and branched sponges.*

*346-347 Sponges and gorgonians sometimes share very similar shapes, produced by the same environmental factors.*

*346 bottom left Rod gorgonians greatly intrigue divers; used to gorgonian sea fans they find these organisms hard to classify.*

*346 bottom right The giant basket star* (Astrophyton muricatum) *is characterized by many branched, mobile arms which by day can closely envelop the gorgonians. At night the arms outstretch to better capture plankton and small organisms carried by the currents.*

The Jewfish *(Epinephelus itajara)*, one of the largest
species, capable of growing to 3 meters in length and
400 kilos in weight, is a symbol of this wealth. At the
other extreme come the highly colored *Gramma
loreto*, just 8 centimeters long, but easily recognized
for their yellow and fuchsia colors that stand out in
front of or inside caves where they should be sought
close to the roof, given their habit of always turning
their belly towards the substratum. The external reef
caves are where glass fish usually station by day,
the shoals acting like cat's eyes at diver's flashlights
and torch. Also stationing in these dark environments
or in the tunnels created by the combined action of
the currents and the corals are squirrelfish and
soldierfish, very similar in color and habits to their
congenerous counterparts in the Indo-Pacific.

*349 bottom
A Caribbean
giant grouper
(Epinephelus
itajara) offers its
large body to a
remora that uses
it both as a prop
and as a source
of food.*

*350 top left
Glass fish abound
in caves or wrecks,
where they hide
hudled together in
the darkest parts.*

*350 top right
A blue chromis
(Chromis cyanea).
This species usually
lives in shoals,
swimming in the
open waters above
the reef.*

*350-351 Snappers
often gather in small
groups in caves on
the sea bottom.*

*351 Wrecks are
always one of the
most fascinating
places to dive at.
Inside a wreck there
is always a whole
new world teeming
with life; the picture
shows a group of
silvery glass fish.*

Although caves, tunnels and cracks are an irresistible draw for divers, offering them the chance to explore veritable labyrinths of coral, the external reef is certainly not lacking in attractions; this is especially so when, past the last plateau of sand, the slope becomes steeper, almost vertical and mingles with the blue at what is called the "drop off", beyond the limits of sports diving. Life is all over these walls and seems to constitute an almost uninterrupted layer, embracing the sea bed and some meters of water, at least as many as are needed for the large tubular sponges or the branches of the corals to reach their maximum dimensions, exceeding

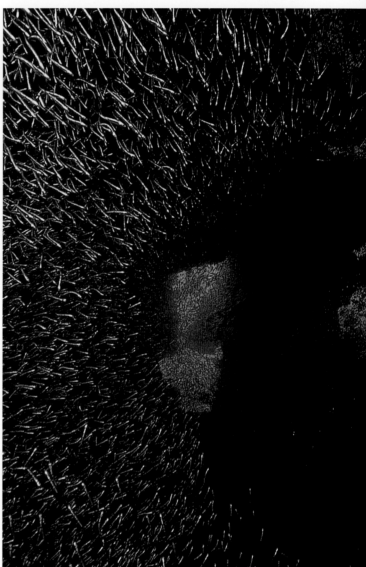

in length or in height those of a diver.
Sticking tightly to the bottom, perpetually in wait, are the hawkfish, totally indifferent to the populations of blue chromis that flit to and fro between the wall and the open water, rhythmically pursuing the plankton moved by the current.

351

352-353
The nurse shark
(Ginglymostoma
cirratum) *can be
photographed
when it stops on
the sea bed by day.*

352 bottom  Nurse
sharks hide in
caves where, taking
advantage of their
lethargy, they can
be seen at close
range. Nonetheless
caution is required
to avoid sudden
attacks.

353 top  Canyons
and crevices in
the reef are where
many snappers
station.

353 bottom
*This lovely close-up
shows the snout of
a queen parrotfish*
(Scarus vetula) *with
its characteristic
green-blue
markings. These*
Scaridae *are fairly
common in the
waters of the
Caribbean.*

*Labridae*, parrotfish, butterflyfish and angelfish dot the whole scenario, busy seeking food or showing off in quiet parades. These seem a speciality of the pairs of French angelfish *(Pomacanthus paru)*, recognized by their dark scales edged with yellow and the surgeonfish, their dark silhouettes creating curtains that can open at any minute to yield the way to some adult male blue parrotfish, easily identified by its large frontal bump. Swimming through the underwater architecture there are opportunities for more interesting encounters.

The tiger groupers *(Mycteroperca tigris)*, for instance, remain in the band of water before the reef and their presence often signals the existence of a cleaning station "run", so to say, by gobies, small *Labridae* or shrimps which could become excellent subjects for underwater photographs.
The sand of the external reef can also offer extraordinary encounters with nurse sharks or with a colony of sand eels, ever ready to withdraw into their dens if a diver come close too quickly and too noisily.

**Great barracuda**
*Sphyraena barracuda*

**Spotted eagle ray**
*Aetobatus narinari*

**Crevalle jack**
*Caranx hyppos*

**Bar jack**
*Carangoides ruber*

**Horse-eye-jack**
*Caranx latus*

**Rainbow fish**
*Elagatis bipinnulata*

**Atlantic spadefish**
*Chaetodipterus faber*

**Cero**
*Scomberomorus regalis*

**Black grouper**
*Mycteroperca bonaci*

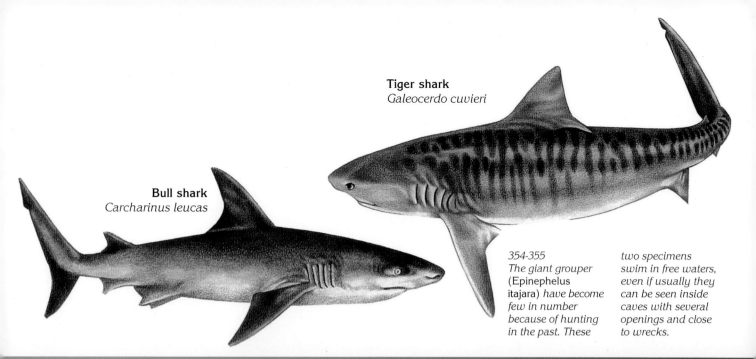

**Bull shark**
*Carcharinus leucas*

**Tiger shark**
*Galeocerdo cuvieri*

*354-355
The giant grouper
(Epinephelus
itajara) have become
few in number
because of hunting
in the past. These*
*two specimens
swim in free waters,
even if usually they
can be seen inside
caves with several
openings and close
to wrecks.*

356-357 *The coastal waters facing the coral reef walls are habitually frequented by spotted eagle rays* (Aetobatus narinari), *easily identified by their spotted back.*

356 bottom
*Sharks are, without doubt, one of the most fascinating attractions of a dive in the Caribbean. They will generally stay away from divers unless intentionally drawn with offers of food made by expert instructors.*

356

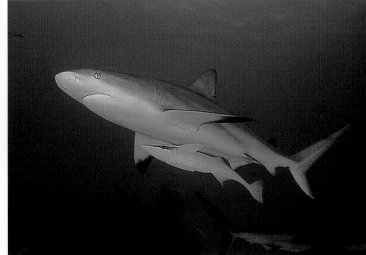

*357 top Sharks and remoras are alway associated. The former offer protection, the latter clean these fearful predators of their parasites.*

*357 bottom Pelagic fish are rarely sighted. The ceros (Scomberomorus regalis), however, before disappearing into the blue will nearly always head for divers, perhaps intrigued by these strange presences.*

The waters fronting the coral waters stretch without apparent limits towards the sea and are, as always, populated by the largest species, their robust bodies permitting them to swim rapidly and powerfully; this is essential to survive in this environment where all are predators.
Fascinating inhabitants though not always welcome are the sharks, not everywhere represented by tranquil grey reef sharks willing to be approached or to perform in exchange for a piece of fish. Tiger sharks and sandbarsharks are by no means rare and their presence can serve as a good deterrent against visits to certain areas without the aid of expert guides.
Free waters do not for this lose their fascination

and diving directly down without following the albeit splendid coral walls means following close, almost sharing the same sensations, on the flight of some formation of eagle rays perfectly recognizable for their lozenge shape, white-spotted wings and bodies and their long tail.
Equally easy to identify are the large barracuda. Mainly solitary, these silvery predators station sneakily in the water and only the half open jaws and sharp teeth give a clue to the formidable predator small fish are faced with. Divers, on the other hand, have on several occasions been approached spontaneously by curious barracuda who, in some areas, have become stable and habitual companions of every dive.

*358 A solitary barracuda (Sphyraena sp.) swims close to a coral bed which is, perhaps, part of its hunting territory.*

*358-359 Carangids (Caranx sp.) are gregarious fish that usually station in front of reef walls. Their social behaviour serves both for hunting and as a defence against larger predators.*

Improbable victims of the barracuda are the jacks, for their size and swimming ability as well as for the habit of living in shoals; as a group they attack the populations of blue-fish (sardines, anchovies) coordinating their attacks with a strategy similar to that of wolves. More prolonged observation, the fruit of repeated dives, will lead to the discovery of other species typical of coastal waters, such as the rainbow fish *(Elagatis bipinnulata)* that, when young, have the strange habit of remaining on the surface in the shade of floating objects or accompanying sharks, mingling with pilot fish. Nearer the coast, but capable of forming shoals of as many as 500 are the Caribbean Atlantic spadefish *(Chaetodipterus faber),* their silvery bodies adorned with wide vertical dark bands and with a compressed,

almost discoidal, silhouette accentuated all the more by the high dorsal and anal fins with backward-curving rays.

Finally, as you return towards the reef, the coastal waters will reserve a last surprise and offer the chance to encounter some black groupers *(Mycteroperca bonaci),* a species capable of greatly exceeding a meter in length and with strange behaviour. Used to considering groupers typical inhabitants of reefs, lovers of caves and shelter, you can only be astounded at the sight of these fish which prefer to remain immobile at a certain distance from the reef, allowing themselves to drift with the currents and moving their fins just the necessary to remain at a safe distance, although they are quite visible in the crystal-clear waters.

359 bottom
*The Atlantic spadefish* (Chaetodipterus faber) *resembles the angelfish, especially when, as in this case, it is so close to the reef. Actually, this species is generally found in open waters.*

359

# THE GORGONIAN FORESTS

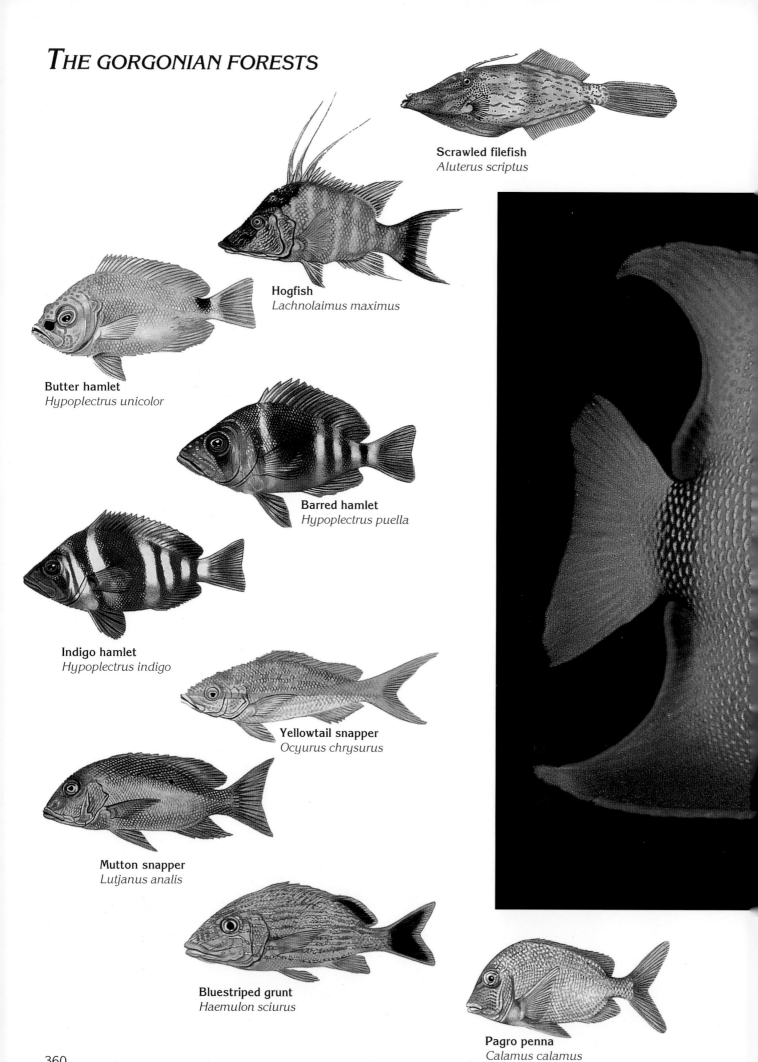

**Scrawled filefish**
*Aluterus scriptus*

**Hogfish**
*Lachnolaimus maximus*

**Butter hamlet**
*Hypoplectrus unicolor*

**Barred hamlet**
*Hypoplectrus puella*

**Indigo hamlet**
*Hypoplectrus indigo*

**Yellowtail snapper**
*Ocyurus chrysurus*

**Mutton snapper**
*Lutjanus analis*

**Bluestriped grunt**
*Haemulon sciurus*

**Pagro penna**
*Calamus calamus*

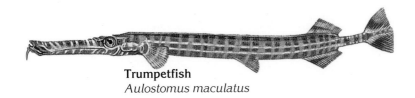

**Trumpetfish**
*Aulostomus maculatus*

**Variegated trunkfish**
*Lactophrys quadricornis*

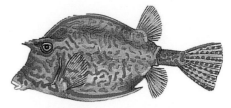

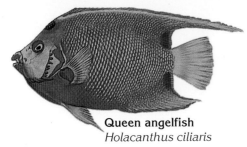

**Queen angelfish**
*Holacanthus ciliaris*

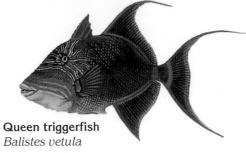

**Queen triggerfish**
*Balistes vetula*

*360-361*
*The* Holacanthus cilarus *well deserves its name of queen angelfish. Its colors make it one of the most striking characters in the gorgonian forests.*

362-363 *The branches of the large Caribbean gorgonian fan colonies are so dense that they create an almost continuous obstacle to the passage of light.*

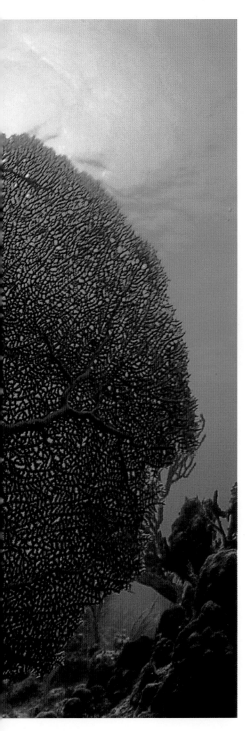

Despite being abundant everywhere and an integrating part of the reef, the Caribbean gorgonians, more than 50 species, deserve a special mention. In some areas they are so dense as to hide all other organisms and constitute veritable forests.

The distribution of these organisms is influenced by a few but essential criteria: the existence of hard sea beds on which the larvae can settle, the presence of constant currents bringing oxygen and food and, lastly, suitable illumination.

The existence of these factors can be easily verified during any dive near gorgonians whose colonies permit the diver to guess the presence of currents and establish the prevailing direction.

*362 bottom Despite its beauty, no underwater photograph can convey the thrill of seeing the gorgonians swaying in the current.*

*363 The picture shows a colony of sea plumes (Pseudopterogorgia sp.) stretching upwards like a huge, undulating ostrich feather.*

*364 top  Some parts of the Caribbean sea beds have become forests with a lush growth of countless different gorgonians.*

*364 bottom  Some gorgonians lose their own identity and become ideal substrata for encrusting species (truncates, sponges, serpulids), adding color to the underwater world.*

*364-365  This massive old hard coral colony has been completely invaded and suffocated by gorgonians and other organisms which have created an isolated microcosm full of life.*

Those coming from the Mediterranean or who have visited the Indo-Pacific will immediately recognize the fan shape of the most characteristic gorgonians that occupy different spaces according to their needs. So the *Gorgonia flabellum* or Venus fan, yellow or greenish-yellow, grows mainly near the fringing band where the hydrodynamism is most accentuated.

In contrast, the *Gorgonia ventalina*, the most widespread and largest species, capable of exceeding 2 meters in height, prefers deeper waters, with less violent but more constant currents. Growing at depths of more than 20 meters, where the waters are clear and the walls of the reef become vertical, are the reddish gorgonians of the depths *(Ilicigorgia schrammi)*. The fans of this species grow profusely in the favourable areas creating forests that conceal the entrance to underwater caves and tunnels.

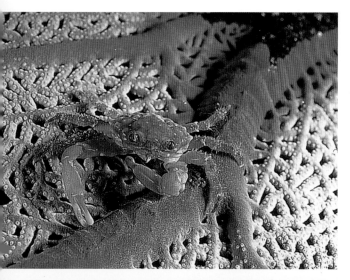

As well as these regular forms, some Caribbean gorgonians take on a different appearance. Certain species have large finger-like candelabra branches that appear extraordinarily fluffy and increase in size when the polyps are expanded. Others have branches with an almost square section that rise vertically from a common base. Very striking are the sea plumes (Pseudopterogorgia sp.), strange gorgonians up to 2 meters in height that resemble ostrich feathers. Before a similar array of species and form it is not surprising that the gorgonians also play host to an unusual and sometimes exclusive fauna. On their branches, and on these alone, move the flamingo tongue molluscs (Ciphoma sp.) in search of polyps, or large oysters and crinoids settle.

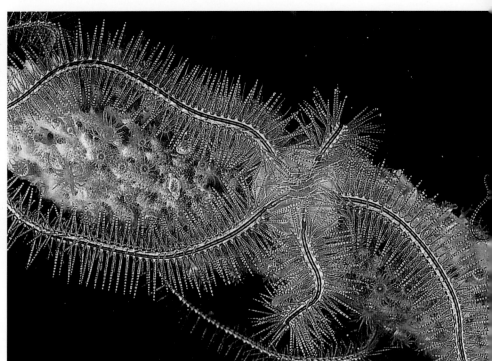

The invertebrates are joined by the numerous fish that populate the gorgonian undergrowth. Some, like the filefish *(Aluterus scriptus)*, seek polyps to feed on, others like the trumpetfish, exploit the sea fans as camouflage and for ambushes. Branches and bushes become the center of the territories of the so-called hamlets, fish so similar to each other and coming in such a wide spectrum of color and shade, between white and black, as to justify the disagreement that still exists between scholars trying to decide whether they are several species or just one with many varieties.

Whatever the solution to the mystery, these fish are ideal subjects for photographs, especially if observed at dusk when the fish can be caught during courting and reproduction.

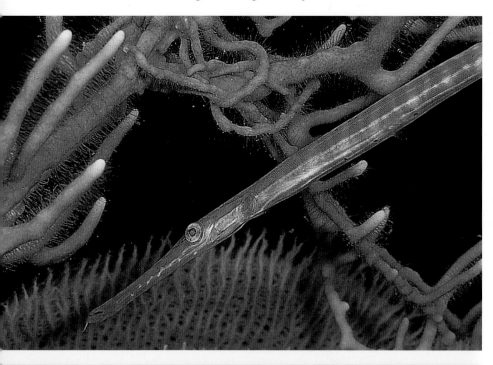

*368-369 Bluestriped grunts* (Haemulon sciurus) *swim slowly through the gorgonians close to the sea bed.*

*369 bottom left Filefish are recognized by their pointed snout and the small first dorsal fin reduced to a sturdy spine.*

*369 bottom right The hogfish* (Lachnolaimus maximum) *is one of the largest and most characteristic* Labridae *in the Caribbean.*

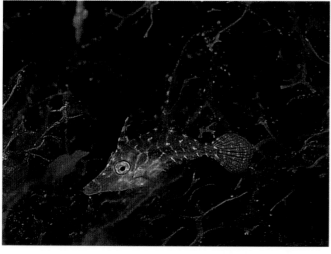

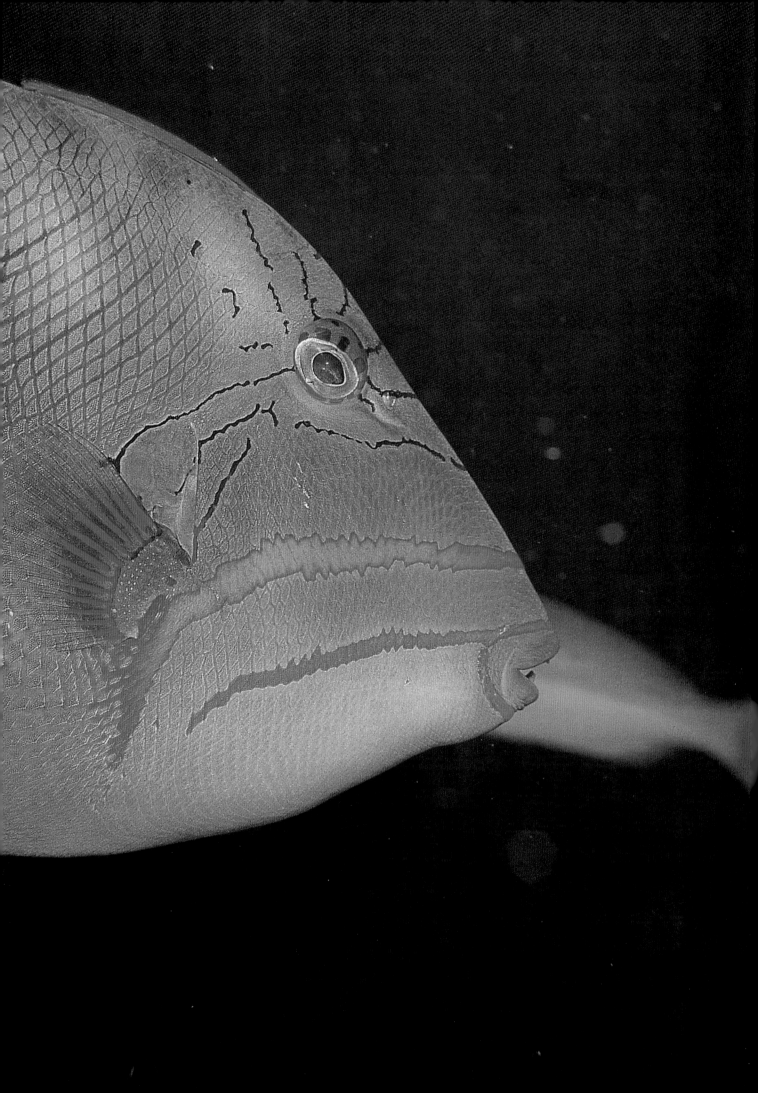

Extraordinarily attractive and colorful
is the queen angelfish *(Holocanthus ciliaris)*
which seems to swim exploiting the currents
that sway the gorgonians.
Equally regal is the *Balistes vetula*, the most
colorful of the triggerfish in the area, always busy
seeking the sea urchins with long spines

it loves to feed on. Practically unmistakable
and very common are the yellow tail snappers
*(Ocyurus chrysurus)* that move in small groups
or solitary just above the gorgonians, ready to
dart towards the bottom as soon as they notice
some rash shellfish destined to precede those
that will be caught by night with far greater effort.

*370  A blue-striped
snout is the main
feature of this
queen triggerfish
(Balistes vetula),
very hard to
approach unless
it is willing.*

*371 top
The cowfish
(Lactophrys
quadricornis) has
large front spines
above its eyes.*

*371 bottom
The yellowtail
snapper (Ocyurus
chrysurus) is a
characteristic
presence in the
waters above the
reef. Despite being
a fast swimmer,
or perhaps for this
very reason,
it approaches,
notoriously slower,
divers without fear.*

# THE SPONGES

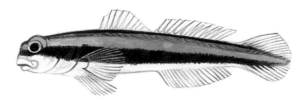

Neon goby
*Gobiosoma oceanops*

Black durgon
*Melichthys niger*

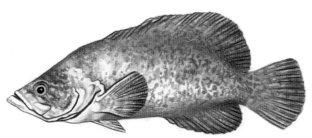

Greater soapfish
*Ryptycus saponaceus*

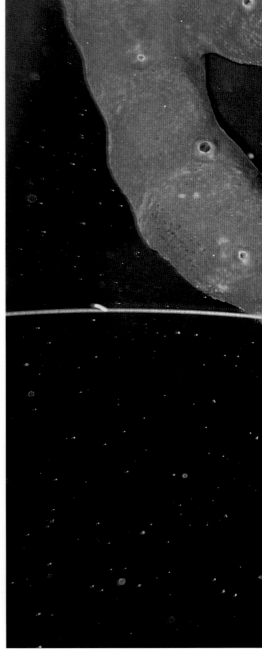

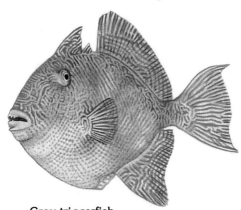

Gray triggerfish
*Balistes capriscus*

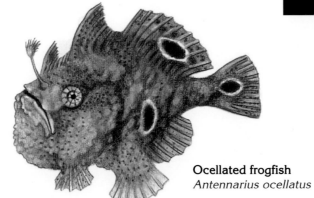

Ocellated frogfish
*Antennarius ocellatus*

**Cleaner goby**
*Gobiosoma qenie*

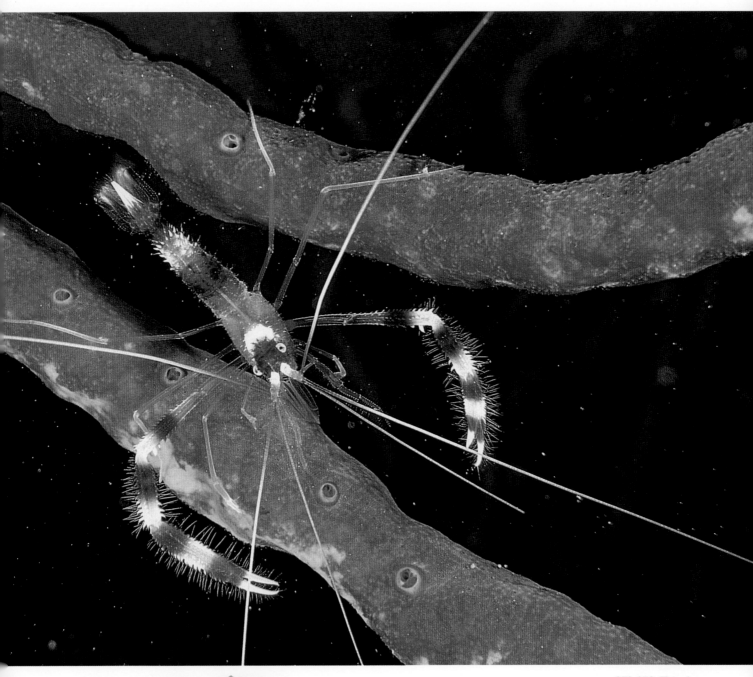

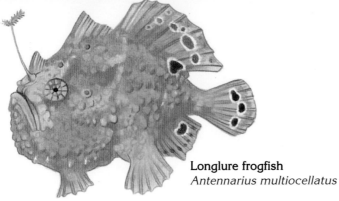

**Longlure frogfish**
*Antennarius multiocellatus*

*372-373 This cleaner shrimp (Stenopus hispidus) in full view on a sponge waves its long white antennae to attract other fish.*

375 top left
These branching
tube sponges
(Pseudoceratina
crassa) *are quite
common, their
bright colors
certainly not
letting them pass
unnoticed.*

375 top right
Along vertical
walls, tube sponges
grow perpendicular
to the bottom,
projecting
themselves towards
the open waters.

375 bottom A huge
orange elephant
ear sponge (Agelas
clathrodes) *rises
from the bottom
with a structure by
no means inferior to
that of hard corals.*

374 Those who
call the Caribbean
"the realm of
sponges" are
merely stating fact.

No description of the Caribbean sea beds
would be complete without reference to the colorful
world of sponges.
Generally underestimated and disregarded except
by experts, these creatures have great importance
in the Caribbean, both for the number of species
(approximately 100) and, above all, for the huge
variety of form and color they present; together
with gorgonians they play the same role as soft
corals in the other tropical oceans.
Even though some species, for instance the large
*Spheciospongia vesparium* sponges, dark in color,
and with a convolute surface as hard as leather,
develop on the sand inside lagoons, most sponges
grow along the external reef, where they compete
successfully with corals for the conquest of space.
During a dive it is not unusual to encounter the
gigantic *Xestospongia muta* barrel sponges,
as hard as stones and almost two meters high;
a diver could hide inside.

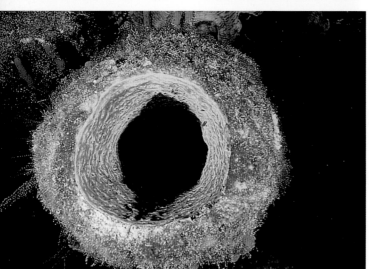

*376 top left  This rock spike rising from the sea bed has become the ideal substratum for species of differently shaped and colored sponges.*

*376 center left The photo reveals the interior of the sponge with all the ducts through which the water absorbed by the external pores passes.*

*376 bottom left At the center of these tube sponges is the pink patch of the Azure vase sponge (Callyspongia plicifera).*

*376 right Branching sponges are somewhat reminiscent of large corals, both for their form and colors, mostly brighter than those of the hard corals.*

*377  These column sponges can grow to far more than a meter in height.*

Equally spectacular are the long tubular sponges, purplish or orange, of the *Aplysina genus*; depending on the case, these may resemble trunks or organ pipes, this form also being adopted by the *Agelas oroides*. These strange shapes, the product of special processes of adaptation to the environment, will not fail to attract the diver's attention, but the same applies to the colors; yellow, fuchsia and brown abound as well as a spectrum of reds and the blues and fluorescent pink of the *Callyspongia plicififera,* a vase-shaped sponge that is one of the most striking and much-photographed in the Caribbean. Not all these creatures are of interest for their form and color. Some are of note because dangerous, as is demonstrated by the *Neofibularia nolitangere,* a scarlet sponge that stands out in the torchlight. Its surface is similar to felt and, if touched, causes a violent and painful allergic reaction with reddening that may last days.

378-379 Giant barrel sponges (Xestospongia muta) *sometimes play host to fish and corals.*

*379 bottom*
*A netted barrel*
*sponge* (Verongula
gigantea) *offers*
*shelter to a queen*
*angelfish.*

*380 top*
*This crab, called decorator because of its unusual habits, is few centimeters long and tries to hide itself on a pink sponge.*

*380 center*
*The yellowline arrow crab (Stenorhynchus seticornis), five centimeters long at the most, displays its long rostrum.*

*380 bottom  The long white antennae of the cleaner shrimp are always in movement and are often the only part of the creature visible when it is hiding in crevices.*

*380-381  This small and curious goby surprised on a sponge remains immobile, relying on its dimensions and markings to save it from the enemy.*

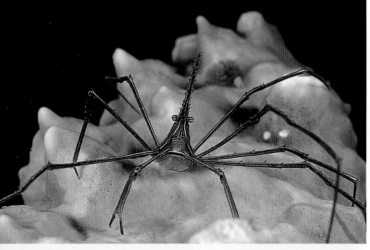

Swimming close to sponges, as too near corals, leads divers to explore a new world made also of noises. It has been discovered that many sponges are inhabited by dozens of small snapping shrimps capable of snapping their chelae, producing a loud crack that fills the water with repeated clicking sounds.

As well as these shrimps the large sponges offer refuge to crabs, cleaner shrimps and young lobsters, without counting the worms, antozoa and many fish, mainly gobies and blennies, that inhabit the pores of the sponges.

382 top  The crown of tentacles of a sedentary worm of the Sabellidae family appears above a branched sponge. Only the tentacles can be seen, the rest of the body being enveloped in the sponge.

382 bottom This frogfish (Antennarius multiocellatus) has a bright yellow coloring in slight contrast with its camouflage habits.

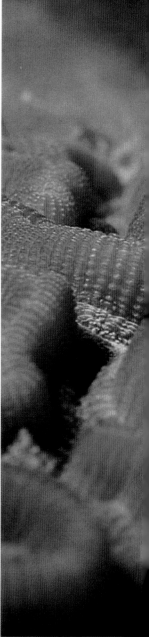

382-383  A small goby (Gobiosoma evelynae) shows off its bright yellow striping as it waits on a star coral for some fish to come along to be cleaned of its parasites.

383 bottom The black spot hides a frightening danger for many small creatures. This is a toadfish camouflaged and about to draw its prey.

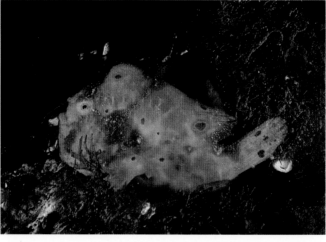

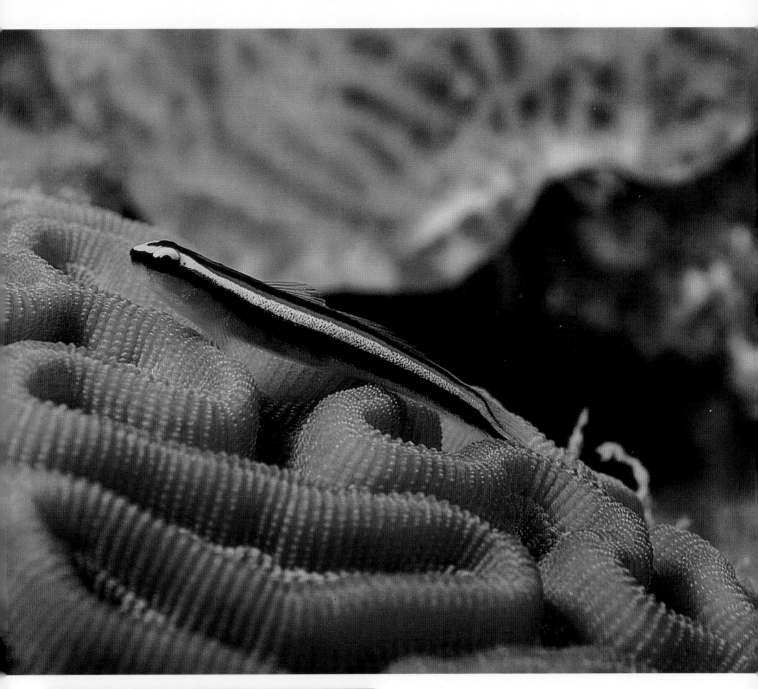

Close to the sponges can be observed considerably larger fish although these are not exclusive to sponges. Often found living on these, well-camouflaged, are frogfish *(Antennarius sp.)*, a cross between scorpionfish and anglerfish, which lie in wait, attracting their prey with their long mobile peduncle above their mouths.
Far more visible are the large grey and the black triggerfish that swim attentive, inspecting the cavities of the sponges in search of prey; this is certainly not the fate of the greater soap fish *(Rypticus saponaceus)* thanks in part to its size but also for the immunity guaranteed by the toxic substances their skin produces, capable of stunning or even poisoning other fish.

# Introduction

All photos are by Kurt Amsler,
except for page 12-13 by Vincenzo Paolillo.

# The Red Sea

Illustration credits:

Marcello Bertinetti / Archivio White Star: Pages 28 bottom, 32 top, 68 top right.

Kurt Amsler: Pages 22-23, 28-29, 31 top, 36 bottom, 47, 55 top, 62-63, 97top, 97 centre right, 100 bottom right, 104-105.

Franco Banfi: Pages 20, 48 bottom, 56-57, 99.

Marco Bosco: Pages 17, 61, 67 top, 84 top.

Carlo De Fabianis: Page 87 centre.

Eleonora De Sabata: Pages 45 top, 100 bottom left.

Pierfranco Dilenge: Pages 26-27.

DUBA: Pages 37 top right, 52, 87 top.

Andrea and Antonella Ferrari: Pages 25, 30 bottom, 32 centre, 40-41, 41 top, 42-43, 52-53, 56 bottom, 58 top, 63 bottom, 67 centre, 68 top left, top right, 71 bottom, 80-81, 81 bottom, 82-83, 84 centre, 84 bottom, 84-85, 87 bottom, 90-91, 92 bottom, 97 bottom, 107.

Paolo Fossati: Pages 26 bottom, 32-33, 39, 66-67, 73, 75 centre, 101, 106.

Itamar Grinberg: Pages 30-31, 63 centre, 89.

Gianni Luparia: Page 91 bottom.

Alberto Muro Pelliconi: Pages 14, 34-35, 58 bottom, 58-59, 74-75, 75 top, 97 centre left, 98 bottom.

Napolitano Francesco / DIAFRAMMARE FOTO SUB: Pages 56 top.

Mark Nissen: Page 31 bottom.

Vincenzo Paolillo: Pages 28 top, 28 centre, 33 top, 48 top, 48-49, 55 bottom, 67 bottom, 71 top, 71 centre, 91 top, 92 top, 92 centre, 92-93, 106 bottom.

Sergio Quaglia / DIAFRAMMARE FOTO SUB: Pages 51 bottom, 68 bottom left, 98 centre.

Roberto Rinaldi: Pages 16, 36 top, 44-45, 45 bottom, 50 bottom, 63 top, 70, 77, 77 top, 98 top, 102-103, 105.

Jeff Rotman: Pages 37 top left, 50-51, 69, 78-79, 81 top, 94-95, 100 top, 106 centre right.

Alberto Siliotti: Page 18.

Alberto Vanzo: Page 19.

Claudio Ziraldo: Pages 21, 37, 41 bottom, 45 centre, 51 top, 65, 75 bottom, 76, 106 centre left.

# The Maldives

Illustration credits:

Marcello Bertinetti / Archivio White Star: Pages 111 bottom, 142-143.

Kurt Amsler: Pages 110, 111 top, 112, 113, 114.

Franco Banfi: Pages 118 top, 120 centre, 132, 140-141, 153 bottom, 156 left centre, 166-167, 200 top.

Claudio Bertasini: Pages 111 centre, 118 bottom, 121 bottom, 124, 144-145, 153 top, 155 right centre, 157, 165 bottom, 168 top, 171 bottom, 177 top, 179 top, 181 bottom, 182-183, 184 left centre, 191 top and bottom.

Claudio Cangini: Pages 129, 146 bottom right, 163 bottom, 168-169, 176-177, 188-189, 188-189, 192-193, 197 left centre.

Andrea and Antonella Ferrari: Pages 120 bottom, 124-125, 130-131, 133 centre, 134 top, 139 top, 140 bottom right, 142 top, 146-147, 150-151, 153 centre, 155 top, 161 centre, 175, 176 bottom, 177 bottom, 178, 180, 181, 186, 187 top, 190-191, 193 bottom, 197 top, 198-199, 199 centre and bottom, 200 bottom, 201 top.

Paolo Fossati: Pages 109, 117 top, 121 top and centre, 122-123, 127 bottom, 133 top, 134 bottom, 146 bottom left, 148 bottom, 148-149, 155 left centre, 155 bottom, 156 top, 156 bottom, 158-159, 162 centre, 162-163, 164-165, 165 top and centre, 168 centre and bottom, 170 bottom right, 171 top, 172 bottom, 172-173, 173 top and bottom, 185, 191 centre, 193 top and centre, 194-195, 196, 200-201.

Vincenzo Paolillo: Pages 116 bottom, 117 bottom, 119 top and centre, 126-127, 127 top, 128 centre right, 136-137, 140 bottom left, 142 centre and bottom, 147, 186-187, 197 right centre, 197 bottom.

Alberto Muro Pelliconi: Pages 128 top, 135, 148 top, 170 bottom left.

Fabio Picarelli: Pages 119 bottom, 141 bottom, 170-171.

Roberto Rinaldi: Pages 116 top, 128 bottom, 133 bottom, 138 bottom, 141 top and centre, 162 bottom, 173 centre, 186 top.

Alberto Vanzo: Pages 118 centre, 120 top, 148 centre, 149 top, 154, 164 top, 199 top.

Claudio Ziraldo: Pages 128 left centre, 138 top, 138-139, 152-153, 156 right centre, 160-161, 161 top and bottom, 162 top, 184 top, 184 bottom.

# Malaysia

Illustration credits:
All photos are by Antonella and Andrea Ferrari.

# The Caribbean

Illustration credits:

Kurt Amsler: Pages 297, 298 top left, 298 centre bottom left, 298 bottom left, 298 right, 299 centre top and centre, 299 right, 301 bottom, 302 centre right, 302 bottom, 303 top right, 304 left, 305 bottom right and left, 306-307, 309, 310-311, 311 bottom, 312 bottom, 314, 315, 316 top and centre, 316-317, 318 bottom, 320-321, 322-323, 322 bottom, 324, 325 bottom left, 326 bottom right, 328 centre, 328-329, 332, 334 left, 336 top, 336-337, 338, 339, 340 centre and bottom, 340-341, 342-343, 344-345, 346-347, 348 top, 348-349, 350 top, 350-351, 352-353, 353 top, 354-355, 356 bottom, 358, 359, 360-361, 362-363, 363 top, 364 bottom, 364-365, 367 top, 368 bottom, 368-369, 369 bottom right, 371 bottom, 374, 375 top, 376 left, 378, 380 top, 383 bottom.

Franco Banfi: Pages 294, 299 bottom left, 300 bottom, 303 bottom left, 305 top right, 313 bottom right, 319 bottom, 326 bottom left, 334 right, 346 bottom right, 347, 366 top left, 368 top, 370, 372-373, 376 right, 377, 380 centre, 380-381, 382-383.

Eleonora De Sabata: Pages 304 right, 305 centre right, 349 bottom, 351 right.

NASA: Page 296.

Vincenzo Paolillo: Pages 323, 324-325, 328 bottom, 335, 352 bottom.

Alberto Muro Pelliconi: Pages 298 centre top left, 301 top, 305 top left, 312-313, 313 bottom left, 316 bottom, 325 bottom right, 327, 336 bottom, 340 top, 353 bottom, 363 bottom, 375 bottom, 380 bottom.

Fabio Picarelli: Page 303 top left.

Roberto Rinaldi: Pages 300 centre left, 302 centre left, 303 bottom right, 311 top, 312 top, 319 top, 326 top, 328 top, 341 top, 366 top right, 368 bottom left, 378-379, 379 bottom, 382 top and bottom.

Egidio Trainito: Pages 318-319, 348 bottom.

Alberto Vanzo: Pages 300 top and centre right, 302 top, 330-331, 346 bottom left, 356-357, 357 right, 362 bottom, 364 top, 366-367, 367 bottom, 371 top.